BERNARD LE BOVIER DE FONTENELLE

Conversations on the Plurality of Worlds

Frontispiece from *Entretiens sur la Pluralité des Mondes*, Amsterdam, 1719.

Conversations on the Plurality of Worlds

Bernard Le Bovier de Fontenelle

With notes and a critical account
of the author's writings by
Jérôme de Lalande

TRANSLATED BY
Elizabeth Gunning

RICHMOND • TIGER OF THE STRIPE

This edition first published in 2008 by

TIGER OF THE STRIPE

50 Albert Road
Richmond
Surrey TW10 6DP

ISBN 978-1-904799-37-5

Typeset in the United Kingdom
by Tiger of the Stripe

Printed and bound
in the United States of America
and the United Kingdom
by Lightning Source

CONTENTS

List of Illustrations

Foreword to This Edition

THIS NEW EDITION of Bernard de Fontenelle's *Conversations on the Plurality of Worlds* is a multilayered book – a fact which will no doubt exasperate some readers, but which, I am sure, will prove enjoyable to most.

It is based on Elizabeth Gunning's translation, published by the Albion Press as a rather attractive duodecimo, with a text area of barely 80 mm × 42 mm, in 1808. This was itself a revised version of Miss Gunning's 1803 edition. Her translation was based on Jérôme de Lalande's 1800 revision of Fontenelle's original 1686 text, as augmented by the author in 1687 and revised by him several more times. To this I have added another layer in the form of notes and foreword. Miss Gunning's spelling and punctuation have been left largely intact, except where they compromise intelligibility, but some glaring typographical errors (often existing in both the 1803 and 1808 editions)[1] have been corrected. Some personal names have been changed to more familiar forms – for instance, this edition has Jérôme de Lalande, where Gunning has de la Lande (and even de le Lande). The form de Lalande is the most common, although the Bibliothèque Nationale de France uses de La Lande.

I have retained the 1808 edition's lack of quotation marks in dialogue. To have inserted them would have added nothing to intelligibility and would have been visually disruptive.

As well as this Foreword, I have added a few endnotes to this edition, indicated by superscript numbers ([1, 2, 3] as opposed to Lalande's footnotes, marked with *, †, etc.). Some may think it fortunate that technical problems prevented me from annotating the footnotes.

ix

Because, above all, this edition is intended to offer an enjoyable and entertaining experience, as Fontenelle's original did, I have added a number of illustrations which are often informative but sometimes largely decorative. Most of the many editions of the *Plurality of Worlds* were illustrated (although often rather badly). In fact the 1808 edition is unusual in containing only one illustration, a portrait of the author.

This edition, therefore, is a rather complex amalgam, an accretion of different editions, for which I make no apology. Read it, enjoy it, peel away the twenty-first-century layer to reveal Gunning's translation. Peel away that to examine Lalande's revision – do we detect a little jealousy of Fontenelle? And, finally, enjoy Fontenelle's text. It may tell you nothing new about astronomy but it will give you fascinating glimpses into seventeenth- and eighteenth-century attitudes to science and to women's education.

Women, Astronomy and Fontenelle

From its first publication in 1686 until the early nineteenth century, innumerable editions of Fontenelle's *Plurality* were published. As well as fourteen Paris editions overseen by Fontenelle himself, not to mention many more French editions issued after his death, there were at least seven Amsterdam editions. There were four attributed English translations: that by 'Sir W. D., Knight' in 1687; John Glanvill's and Aphra Behn's, both of 1688; and Elizabeth Gunning's first published in 1803. Most of these were reprinted many times. Among the other translations was a Greek one by Codrika. Surprisingly, it doesn't seem to have appeared in German until the publication of *Dialogen über die Mehrheit der Welten* in 1780.

Why exactly astronomy came to be seen (as botany was by the Victorians) as a suitable past-time for young ladies is unclear, but

Fontenelle's book certainly played a part in making it popular among the ladies themselves.

Women in the seventeenth and eighteenth centuries were generally excluded from all roles in the sciences, law, the priesthood, the armed forces, politics, often the monarchy, and many other areas of life. It was widely accepted (by men!) that women were not intellectually equipped for these roles. The French philosopher Nicholas Malebranche went as far as to say that

> All that depends upon the Taste falls under their Jurisdiction; but generally they are incapable of Penetrating into Truths that have any Difficulty in the discovery. All things of an abstracted Nature are incomprehensible to them. They cannot imploy their Imagination in disentangling compound and perplexed Questions.[2]

However, he did concede that 'there are some Women to be met with, who have a greater solidity of Mind than some Men.'

Whatever the barriers, there were always women who managed to break into 'forbidden' areas, such as Mary Anne Talbot (cruelly exploited by Captain Essex Bowen, who did much to damage Elizabeth Gunning's reputation – see p. xxvi) who served in the Royal Navy disguised as a man.

Women wishing to study the sciences benefitted from the traditional male prejudice in favour of the classics which relegated science to an inferior status. Women, who were generally denied a classical education, could claim to approach the new disciplines with minds unburdened by the past.[3]

Mathematics and astronomy in particular seem to have attracted women, including a number of really talented scientists. It is doubtful, for instance, if Clairault and Lalande could have completed their calculations on Halley's Comet without the help of Madame Lepaute (p. xx). Émilie du Châtelet (1706–49) not only translated Newton's *Principia Mathematica* into French but also made several original contributions of her own to astronomical physics. Caroline Herschel (1750–1845) made significant contributions to astronomy, not least an amazingly thorough reworking and correction of Flamsteed's star catalogue. She and Mary Somerville (1780–1872) were made the first honorary women members of the Royal Astronomical Society in 1835.

No doubt many other women astronomers have been forgotten because their work was either ignored or passed off as that of their male colleagues. The comet discovered by Maria Kirch (1670–1720), for instance, was named after her husband, not her.

Plurality of Worlds would not, of course, have been intended for these serious astronomers. In fact, they would have known considerably more about astronomy than Fontenelle. His book offers very little serious information about astronomy; it doesn't even give any advice on the use of telescopes, and is severely compromised by Fontenelle's attachment to the idea of Cartesian vortices, a rather clumsy pre-Newtonian model of the universe. To be fair, the first edition was published the year before *Principia Mathematica*. Fontenelle's aims were clearly to entertain and stimulate the imagination, while offering a modest amount of information – a very sound educational approach and one in which he succeeded admirably.

Bernard Le Bovier, Sieur de la Fontenelle

Bernard de Fontenelle was born in Rouen, Normandy, on 11 February 1657. His father's family were mostly lawyers and came originally from Alençon, about ninety miles south-west of Rouen. When his great-grandfather, Nicolas Le Bovier, married Madeleine de Pallue, the foster-sister of Jeanne III of Navarre, the whole family converted from Catholicism to Calvinism, but his son, Isaë, converted back to Catholicism, so Bernard's father, François, and Bernard himself were born into the Catholic faith. Rouen, in any case, was an oasis of religious toleration in an age of bigotry.

Portrait of Bernard de Fontenelle by Louis Galloche.

Fontenelle's mother, Marthe, also came from a family of law-yers, but her brothers, Pierre and Thomas Corneille, were, of course, better known as dramatiſts.

Bernard was a frail and sickly child, and it has been suggeſted that this was the reason for his lifelong fear of extremes of tem-perature and of emotion.[4]

He was educated by Jesuits at the local Collège de Bourbon (now the Lycée Pierre Corneille), where his uncles had also ſtud-ied. Like moſt Jesuit colleges, it offered a high ſtandard of educa-tion, but it was notably deficient in mathematical teaching.

Fontenelle was expeĉted to follow his father into the law, al-though there was also some expeĉtation that he might join the prieſthood. But he soon showed an intereſt in other direĉtions, discovering the works of Euclid and Descartes. Perhaps it was the very fact that they were outside the curriculum that first attracted him, but it was the beginning of a life-long fascination with mathematics and the sciences.

By the age of fourteen or fifteen, he was a regular visitor to his uncle Thomas's house in Les Andelys, about twenty-five miles south-east of Rouen, where he learnt the rudiments of astronomy. In fact, it seems likely that Thomas, who went on to produce a *Dictionnaire des Termes d'Arts et de Sciences* in 1694, introduced his nephew to a wide range of disciplines.

Meanwhile, Fontenelle was naturally being drawn into the literary world. His first published work, *Le Coq*,[5] was printed when he was fifteen. Soon he found himself writing for the *Mercure Gallant*,[6] having been introduced by Thomas Corneille, who was also a contributor. It was almost inevitable that the young man would attempt to emulate the success of his illustrious uncles, particularly Pierre, in writing for the stage, and in 1680 his historical play, *Aspar*, was performed in Paris. Despite much puffing by de Visé and the *Mercure Galant*, the opening night attracted a small audience and it only ran for three performances. He followed this with a comedy, *La Comète*. It was hardly more successful, but some of its astronomical themes were to reappear in the work for which he is mostly remembered.

The failure of these plays must have been a blow to the young writer, but he persevered, and even returned to writing for the stage in 1689 with the opera *Thétis et Pelée* – another disaster, although Voltaire claimed it was performed 'with great success'.[7]

In 1683 Fontenelle published his first really successful work, *Lettres Galantes du Chevalier d'Her…*, a slight but amusing collections of letters,[8] which was reissued in an enlarged form in 1685.

Another, even more important and successful book was published in 1683, his *Nouveaux Dialogues de Morts*. The idea of imagining dialogues between dead people was not at all new, dating

Thomas Corneille.

back at least to the *Νεκρικοί Διάλογοι* (Dialogues of the Dead) of Lucian of Samosata in the second century A.D., but Fontenelle (who acknowledged his debt to Lucian) managed the task with great wit and charm. The dialogues were divided into three parts:

dialogues between ancients, dialogues between ancients and moderns, and dialogues between moderns.[9]

In 1685, Louis XIV, the so-called Sun King, revoked the Edict of Nantes which had, at least in theory, protected Huguenots from persecution and in 1686, Fontenelle published what may be considered his most daring work, although it was only about 800 words. His *Relation sur l'isle de Bornéo* purports to be a letter sent from Batavia (modern Jakarta) in 1684 concerning a civil war in Borneo between the rival queens, Mréo (an anagram of Rome) and Eénegu (allowing for u/v interchangeability, an anagram of Genève). It was a very thinly disguised attack on the feuding between Catholicism and Calvinism in which neither side comes out with much credit. Mréo is a tyrant who wants all her ministers to be eunuchs, while Eénegu claims to be the legitimate heir to the throne which the writer says is 'almost unbelievable'.

However, it is for another work published in 1686 that Fontenelle is remembered, his *Conversations on the Plurality of Worlds*, a book which was regularly reprinted and translated into many diffferent languages for the next 140 years.

Fontenelle did not rest on his laurels. In the same year[10] he published his *Histoire des Oracles,* an adaptation of Anton van Dale's *De Oraculis Veterum Ethnicorum Dissertationes* (1683). As Aphra Behn (who also translated *Plurality of Worlds*) says in the dedication to her English translation,[11]

> The learned and ingenious *Fontenelle* is the Author of this Book I most humbly offer to your Patronage, which as himself confesses in the Preface, he took mostly from the famous *Vandale*, a Dutch-man. The Original was large and tedious: Our Author has left out all that might swell the Bulk of his Book, without any mighty Improvement or Pleasure to the Reader, and so has made this perhaps, tho' only a Copy, better than the Original.

Strangely, despite this, Fontenelle is not mentioned on the title page. At this date, the English publisher muſt have thought that van Dale's name had more cachet than Fontenelle's.

The *Hiſtoire des Oracles* was another controversial book because, although it oſtensibly dealt with pagan superſtition, it also contended that oracles did not cease with the birth of Chriſt. There was a clear implication that Chriſtianity itself was, if not based on superſtition, at leaſt tainted by it. Fontenelle's mockery of pagan prieſts may be seen to apply to Chriſtian prieſts as well. The book was clearly interpreted in this light by the Jesuit, Jean François Baltus, who published a *Réponse à l'Hiſtoire des Oracles de M. de Fontenelle* in 1707. That this response had taken over twenty years gives some indication of how hard Baltus had worked to counter this perceived threat to the Church. It was a laborious argument, full of footnotes in Greek, which would have had little impaĉt on those who enjoyed Fontenelle's light touch.

Also published in 1686 was *Doutes sur le Syſtème Physique des Causes Occasionelles* in which Fontenelle criticised the work of Nicolas Malebranche, the mathematician and scientiſt. The arguments were complex, but it seems that Fontenelle's fundamental objeĉtion to Mallebranche's work was his insiſtence that God was inimately involved in the mechanics of the universe.

In 1688, Fontenelle published a book of verse, *Poésies Paſto- rales*, which made so little impaĉt that it did not even make its way into later colleĉted editions of his works. In the same year, however, he also published another controversial work, *Digres- sion sur les Anciens et les modernes*. It was one of the firſt shots fired in the a rather pointless argument, the *querelle des anciens et des modernes*, about the relative merits of literature and other arts founded on classical models and the more forward-thinking approach adopted by the *modernes*. Fontenelle espoused the latter cause, while a number of eſtablished writers such as Racine and Nicolas Boileau-Despréaux supported the cause of the ancients.

Feelings ran high and when Fontenelle applied to join the Académie française a number of the *anciens* tried to block him. Nonetheless, Fontenelle was elected to the Académie in 1691. He was subsequently admitted to membership of the Académie des Inscriptions and the Académie des Sciences of which he became Permanent Secretary in 1697. In this capacity, he produced the *Histoire du Renouvellement de l'Académie Royale des Sciences en M.DC.XCIX* (3 vols, 1708, 1717, 1722) and some sixty-nine *éloges*, biographical memoirs of the members, including one on his uncle Pierre Corneille.

Among Fontenelle's later works were his *Élémens de la Géométrie de l'Infini* (1727) and *Apologie des Tourbillons* (1752). He died on 9 January 1757, just a month short of his one hundredth birthday. He attributed his longevity to eating strawberries.

Joseph Jérôme Lefrançais de Lalande

Jérôme de Lalande was born on 11 July 1732 in Bourg-en-Bresse, in eastern France, nestling at the base of the Jura mountains. Like Fontenelle, he received a Jesuit education and his father intended him to follow a legal career. Like the author of *Plurality of Worlds*, the young Lalande soon became distracted by the allure of the sciences, particularly astronomy.

Later portraits show a strange-looking man with staring eyes and wild hair, a smile which is more unnerving than benevolent playing on his lips; but an earlier, elegant portrait by Fragonard is much less sinister. At any age, he would have attracted attention as he was only 4ft 6 inches (1.37 metres) tall.

Jérôme de Lalande.

There is a certain haughtiness in his 'Critical Account of the Life and Writings of the Author', when he tells us that 'the reputation of Fontenelle renders him respectable, even in his mistakes.' There is surely also a hint of jealousy when he says:

> The *Astronomy for Ladies*, which I have published as a substitute for this book, would be more instructive, but less amusing; therefore as it will be but little read, I shall endeavour to supply the defects of Fontenelle's work; by adding to the original some ideas more exact than his own.

Whether motivated by a genuine desire to educate or by a wish to achieve high book sales, Lalande was pragmatic enough to adapt *Plurality of Worlds* to his purpose.

The reader of his 'Critical Account' and notes would readily get the impression that Lalande was an important man, a great authority on astronomy. Perhaps they would be right.

A brief examination of his life would seem to confirm Lalande as a scientist of the first order. While studying Law in Paris, he attended lectures at the Collège Royale by Nicolas Delisle on astronomy and by Pierre Charles Le Monnier on mathematical physics.

In 1751, the Abbé Nicolas Louis de Lacaille went to the Cape of Good Hope to measure the lunar parallax, with Lalande being deputed by Le Monnier to take simultaneous measurements in Berlin. Lalande and Le Monnier disagreed over the calculations and, when a commission found in his favour, Lalande, with a characterisitic lack of humility, pressed his claim and fell out with Le Monnier. Nonetheless, he was swiftly admitted to both the Prussian Academy and the Académie des Sciences.

Lalande's fame (which he vigorously promoted) was further enhanced by his collaboration with Alexis Clairault and Nicole-Reine Lepaute, wife of Jean André Lepaute, clockmaker to the French king. The team successfully revised Halley's calculations to predict the return of his comet. It may have owed more to Lepaute than to Lalande.[12]

Lalande went on to publish a great many books, including a six-volume work on astronomy and *Voyage d'un François en Italie*, in a staggering eight volumes.[13] He showed himself particularly interested in industries and mechanical processes, such as the manufacture of pasta, and elsewhere he wrote on papermaking. The notes of Lalande's trip to England in 1763 show the same characteristic. As Richard Watkins says:[14]

> Most of Lalande's writing has no *feeling*, no *smell* about it. What did people look like? What did buildings look like? What were the streets like? How did the city sound, was it quiet or noisy? His description of the Tower [of London] is a brief list of facts. His walk with Berthoud (Sunday 22nd May) gives me no images of appearance of this area. Nowhere does he describe the people he met. He simply lists names and places...

More damningly, Watkins observes:[15]

> ... if we examine the technical entries we see that they have two common characteristics. Firstly, they are reports of the statements of other people and there are almost no entries where Lalande expresses his own views. And secondly, they are often rather vague and superficial. The remarks on experiments in fluid friction, electricity and magnetism are fairly obscure, and appear to be the notes of an informed observer and not the analytical writings of a scientist.

Whatever his abilities, Lalande's reputation remained intact, and, indeed, grew throughout Europe, incluuding Britain where he was elected to the Royal Society.

Perhaps the strangest thing about him was his habit of eating spiders.[16] They were clearly not as nutritious as Fontenelle's strawberries as he died in 1807 at the tender age of seventy-four.

Elizabeth Gunning

Elizabeth Gunning (1769–1823) was the daughter of Major General John Gunning of Castle Coote in Ireland and the novelist Susannah Gunning (née Minifie). John Gunning, the son of a barrister, was away fighting in North America for much of Elizabeth's childhood. He 'shewed the greatest proofs of military conduct and personal bravery' at Bunker's Hill and was mentioned in despatches.[17] As will be seen, other aspects of his conduct were more open to criticism.

The General's sisters, Elizabeth (c. 1733–1790) and Maria (c. 1732–1760), were famous society beauties whose good looks enabled them to make much better matches than might have been expected, given their modest wealth and social status. Elizabeth became first Duchess of Hamilton and then (on the death of her first husband) Duchess of Argyle. The notoriously tactless Maria[18] became Countess of Coventry.

The 'luck of the Gunnings' became proverbial in Ireland and it seems from subsequent events that either the young Elizabeth or her mother may have assumed that she would be able to marry as well as her aunts. At all events, Elizabeth seems to have fancied herself in love with her cousin, the Marquess of Lorne, 'a fancy too pleasing to be rejected,' said the Marquess's sister, Lady Charlotte Bury, née Campbell (1775–1861).[19] She adds rather cruelly that her brother, 'in the first éclat of youth, novelty and good looks, had little inclination to think of an ugly cousin at home, but the poor girl chose to imagine otherwise.'[20] *Was* Elizabeth ugly? Lady Bury says 'All the prints of her make her decidedly very plain,'[21] and it is true that an engraving of 1802 after a painting by Robert Ker Porter does not make her appear particularly good-looking. On the other hand, the stipple engraving of 1796 by Bartolozzi, after 'Saunders Jr.',[22] is really very pretty. It might be objected that Bartolozzi's engravings often flattered the sitter, but even Gillray's *Siege of Blenheim* makes her look attractive,

Elizabeth Gunning, the 'ugly cousin'.
A stipple engraving of 1796 by Francesco Bartolozzi,
after a painting by Saunders.
Courtesy of New York Public Library.

The seige of Blenheim — or — the new system of GUNNING, discover'd' by William Gillray, 5 March 1791. Elizabeth Gunning is sloping off, straddling the canon. On the left General Gunning straddles the canon. On the right the Duchess of Bedford offers shelter to Elizabeth and her mother.[23]

Reproduced by courtesy of the Wellcome Library.

and Gillray was more than happy to make his subjects hideous –
witness his treatment of her mother and the Duchess of Bedford
in the same print.

Rumours soon circulated that Elizabeth was to marry the young
Marquess. However, another rumour started to circulate in 1790 –
that Elizabeth was to marry the Marquess of Blandford, heir to the
Duke of Marlborough and an even better match. It was said that:

> … about two years ago, the Marquis of Blandford met the accom-
> plished Miss Gunning at a Ball, and had the good fortune to engage
> the lady's hand as his partner for the evening. Soon after, Miss Gun-
> ning received a letter from the Marquis, as she believed, expressing
> very tender sentiments of admiration, and soliciting permission to
> visit and to correspond with her… the correspondence went on for
> some months, until the Duke of Argyle suggested some doubts of the
> Duke of Marlborough's being acquainted with the affair.[24]

The *Gazeteer and New Daily Advertiser* declared:

> Miss Gunning will soon be a Marchioness, She has *shot* the Marquis
> of Blandford through his heart, and she is to wear it as a *trophy*.[25]

It was said that General Gunning had written to the Duke of
Marlborough and received back a letter confirming that 'an alli-
ance with the ancient family of the General would be highly desirable.'[26] However, the letters were soon denounced as forgeries,
although the culprit and motive remained obscure:

> The fact is, that deep, dark, and mysterious as the plot has been, it
> will turn out to be an artful machination in a quarter from which
> the young Lady should rather have received protection than injury,
> practised for the purpose of drawing off the affections of a young
> Nobleman really enamoured of her charms, and to whose passion
> they were adverse.

> Miss Gunning was, equally, with her mother, the dupe of the contriv-
> ance…[27]

General Gunning clearly wished the world to believe that he was an innocent dupe of his wife and daughter, and he banished them from his house. In a ſtrange twiſt, *The Times* of 23 February 1791 tells us that the Duchess of Bedford, Blandford's grandmother, provided Elizabeth and her mother with a house in Pall Mall. In the same article, *The Times* remarked sourly:

> Miss Gunning, though handsome, is not so much celebrated for her beauty, as for her accomplished manners and amiable disposition. – It is therefore, with great regret, we observe, that vanity have [sic] so far triumphed over discretion, as to have given rise to the many remarks which naturally occur on a late transaction.

Mrs Gunning, in an open letter to the Duke of Argyle, claimed that the whole affair had been the work of the General's cousin Essex Bowen[28] and his wife. Bowen implicated Elizabeth in a circumſtantial, if implausible, account. Moſt damningly, he claimed that the letter which the General sent to the Duke of Marlborough was intercepted by Elizabeth who then arranged for a forged reply to be sent to her father.

There is some reason to miſtruſt Bowen's account, not leaſt because Gunning himself later admitted that he had known of the affair:

> … I found it necessary to make some grand efforts to extricate myself from the business, and preserve my reputation unsullied.

> For this, I summoned a grand council of the whole family; and it was unanimously agreed upon, that I should pretend ignorance on the occasion – look upon the Marquis of B———'s imaginary proposals as having actually taken place – and write to the Duke of M——— for an explanation of his son's conduct. We easily foresaw that an ecclaircissement of this nature would open the eyes of the world, and occasion a total revolution in the family. But this we were content to bear – it was only for me to renounce, *in appearance,* and keep the secret on which all the politics of our little commonwealth depended.[29]

Bizarrely, a year after he sacrificed his wife's and daughter's reputations to preserve his own, Gunning published an account in which he not only confessed his part in this affair but also admitted to being a compulsive philanderer whose conquests included two duchesses, fourteen countesses, four viscountesses, seven baronesses, thirteen baronets' ladies, and many others.[30] It is perhaps a reflection of Gunning's obsession with social standing that he should list them thus.

If the General is to be believed (and it is hard to see what he has to gain by lying at this stage), Elizabeth played Blandford and Lorne against each other, dithering as to which she preferred. When she finally decided on Lorne, his ardour had cooled, so Mrs Gunning forged a few letters from Blandford in the hope of rekindling Lorne's interest.

Whoever was to blame, Elizabeth lost both suitors and her reputation. Moreover, her father fled the country with his mistress, Mrs Duberly, to escape *his* creditors and *her* husband. It is no wonder that she followed her mother into a writing career; she probably needed the financial security. Her first novel, *The Packet*, was published in 1794. No doubt, as she anticipated, the notoriety of her name improved the sales. She went on to publish five other novels[31] and a translation of the *Memoirs of Madame de Barneveldt* before translating *Conversations on the Plurality of Worlds* in 1803. In that year, at thirty-four, she married a Major J. Plunkett who had been involved in the Irish rebellion of 1798 and condemned to death. He had managed to escape to France and was later allowed to return to England, thanks to the intervention of the Duke of Argyle.[32]

Elizabeth wrote three more novels, *The War-Office* (1803), *The Exile of Erin* (1808) and *The Victims of Seduction or, Memoirs of a Man of Fashion* (1815)s as well as several translations and some children's books. She died in Long Melford on 20 July 1823, the death notices not even mentioning her maiden name, let alone her previous notoriety. Perhaps, indeed, her reputation had never been too badly damaged. At any rate, she was allowed to dedicate *The War-Office* to the Duke of York and *The Victims of Seduction* to the Princess of Wales.

Her novels are not much admired by modern critics. They do not pretend to be anything but popular romantic fiction but they are reasonably well written examples of the genre. Perhaps this new edition of her translation of *Conversations on the Plurality of Worlds* will do something to redeem her reputation. The main credit for the popularity of this book over the centuries must, of course, lie not with her or Lalande but with Bernard de Fontenelle. Nonetheless, Elizabeth Gunning's translation successfully captures the charm and humour of the original in a way that some other translations have failed to do.

PMD
November 2008

Critical Account
of the Life and Writings of the
Author

BY JÉRÔME DE LALANDE

WHENEVER I HAVE entered into conversation with any sensible woman on astronomy, I have always found that she had read Fontenelle's *Plurality of Worlds;* and that his book had excited her curiosity on the subject. As it has been so much read already it must continue to engage attention: I therefore thought it would be useful to point out its faults; to add some observations, without which the reader would be led into error with respect to the vortices; to make known the late discoveries; and to shew what numbers, before our author, had written on the plurality of worlds. But I have made no alterations in the text; the reputation of Fontenelle renders him respectable, even in his mistakes.

The Astronomy for Ladies,[33] which I have published as a substitute for this book, would be more instructive, but less amusing; therefore as it will be but little read, I shall endeavour to supply the defects of Fontenelle's work; by adding to the original some ideas more exact than his own.

M. Codrika[34] has translated it into Greek, with explanations taken from my *Astronomy.*

M. Bode[35] has translated it into German; and his translation has already gone through three editions: the last is that of 1798, Berlin, in octavo, Bernard de Fontenelle, *Dialogen über die Mehrheit der Welten.*

When Voltaire published, in 1738, his *Essays on the Elements of Newton,* he began with these words: 'Here is no Marchioness; no imaginary philosophy.' It was supposed that he here alluded to Fontenelle; this he contradicts by saying: 'so far from having this

1

book in view, I publicly declare that I consider it one of the best works that ever was written' (*Mem. de Trublet,* p. 135).

This book has been printed a hundred times; the handsome edition of Fontenelle's *Works,*[*] in folio, published at the Hague in 1728, with figures by Bernard Pickart; the still more beautiful edition of the *Worlds* alone, edited by Didot the younger, in 1797, in folio, are master-pieces of typography; but in them nothing is found but the original work; therefore I consider our edition far preferable.

I shall here give a short account of the author of this work.

Bernard le Bovier[†] de Fontenelle was born at Rouen, February 11, 1657. He died January 9, 1757.

The first efforts of his genius were directed to poetry: at the age of thirteen he had composed a Latin poem: about the year 1683 he devoted himself to literature and philosophy. In 1699 he began *l'Historie de l'Académie des Sciences,* which he continued with great success during forty-two years. Few persons have contributed more to the progress of the sciences than he has done, by accommodating them to every capacity, and inspiring by his panegyrics, a love of study. For my part, I feel a pleasure in acknowledging that I am indebted to him for the germ of that insatiable activity of mind I have experienced ever since the age of sixteen. I could find nothing in the world like the Academy of Sciences, and ardently wished for the happiness of seeing it, long before I had any idea of the possibility of one day belonging to it.

In 1757 he published his *Elements de la Géométrie de l'infini;* this was merely the amusement of a man of genius who had heard a little of geometry, and chose to hazard his opinions on the subject.

We may find an eulogium on our author in *l'Historie de l'Académie des Sciences* for 1757, in the *Mémoires de l'Académie des Belles Lettres,* and in a work written entirely on the subject, published by

---

*   That edition does not contain the account of the bees, which is in the present edition.

†   Lebeau writes the name Le Bouyer, from the family name, in the *Memoirs de l'Académie des Inscriptions;* but it is pronounced le Bovier. (*Mem.* p. 19.)

Trublet[36] in the year 1761, entitled *Mémoires pour servir à l'Histoire de la vie et des Ouvrages de Fontenelle*. In these memoirs a particular critique shews us the various merits of Fontenelle's works: there is also an article by Trublet in the edition of Moréri,[37] published in 1759.

I have remarked in the twentieth book of my *Astronomy*, that in every period of time it has been believed that the planets were inhabited, on account of their resemblance to the earth. The idea of the plurality of worlds is expressed in the *Orphics*, those ancient Grecian poems attributed to Orpheus (Plut. *de Placitis Philosoph. l.* 2, cap. 13). Proclus has preserved some verses in which we find that the writer of the Orphics places mountains, men, and cities in the moon. The Pythagoreans, such as Philolaüs, Hicetas, Heraclides, taught that the stars were all worlds. Several ancient philosophers even admitted an infinity of worlds beyond the reach of our sight. Epicurus, Lucretius and all the Epicureans were of the same opinion; and Metrodorus thought it as absurd to imagine but one world in the immensity of space, as to say that only one ear of corn could grow in a great extent of country. Zeno of Eleusis, Anaximenes, Anaximander, Leucippus, Democritus, asserted the same thing: in short there were some philosophers who, although they did not consider the rest of the planets inhabited, placed inhabitants in the moon; such were Anaxagorus, Xenophanes, Lucian, Plutarch, (*De Oracular. defectu. De Facie in orbe Lunæ*), Eusebius, Stobius. We may see a long list of the ancients who have treated on the subject, in Fabricius, (*Biblio. Græca*, t. 1. cap. 20.) and in the *Mémoire de Bonamy* (*Acad. des Inscriptions*, tom. ix.) Hevelius[38] appeared as firmly persuaded of this opinion in 1647, when he talked of the difference between the inhabitants of the two hemispheres of the moon: he calls them *selenitæ*, and examines at length all the phœnomena observed in their planet, after the example of Kepler (*Astron. Lunaris*). It was maintained at Oxford, in certain themes which are mentioned in

Hevelius's map of the moon, 1647.

the *News of the Republic of Letters,*[39] June 1764, that the system of
Pythagoras on the inhabitants of the moon was well founded; two
years afterwards Fontenelle discussed this subject in his agreeable
work. There are farther details of the different astronomical opinions at the end of Gregory's book. For the objections, we may
refer to Riccioli. (*Almagestum* tom. 1, p. 188, 204). In 1686 the
Plurality of Worlds was adorned by Fontenelle with all the beauties of which a philosophical work was susceptible. Huygens (who
died in 1695) in his book entitled *Cosmotheoros,* published in 1698,
likewise enters largely into the subject.

The resemblance between the earth and the other planets is so
striking, that if we allow the earth to have been formed for habita-

tion, we cannot deny that the planets were made for the same purpose; for if there is, in the nature of things, a connection between the earth and the men who inhabit it, a similar connection must exist between the planets and beings who inhabit them.

We see six planets around the sun, the earth is the third; they all move in elliptical orbits; they have all a rotatory motion like the earth, as well as spots, irregularities, mountains; some of them have satellites; the earth has one satellite: Jupiter is flattened like our world; in short there is every possible resemblance between the planets and the earth: is it, then, rational to suppose the existence of living and thinking beings is confined to the earth? From what is such a privilege derived but the grovelling minds of persons who can never rise above the objects of their immediate sensations?

Lambert[40] believed that even the comets were inhabited (*Système du Monde*, Bouillon; 1770). Buffon[41] determines the period when each planet became habitable, and when it will cease to be so, from this refrigeration (*Supplémens*, in 4to. tom. 11). What I have said of planets that turn round the sun, will naturally extend to all the planetary systems which environ the fixed stars; every star being an immoveable and luminous body, having light in itself, may properly be compared with our sun. We must conclude that if our sun serves to attract and enlighten the planets which surround it, the fixed stars have the same use. It is thought that the sun and fixed stars are uninhabitable because they are composed of fire; yet M. Knight,[42] in a work written to explain all the phenomena of nature, by attraction and repulsion, endeavours to prove that the sun and stars may be habitable worlds, and that the people in them may possibly suffer from extreme cold. M. Herschel[43] likewise thinks the sun is inhabited (*Philos. Trans.* 179. p. 155, *et suiv*).

Some timid, superstitious writers have reprobated this system, as contrary to religion: they little knew how to promote the glory of their Creator. If the immensity of his works announce his power, can any idea be more calculated than this to exhibit their magnificence and sublimity? We see with the naked eye, several thousands of stars; in every part of the firmament we discover with telescopes, innumerable others; with more perfect telescopes, we still find a multitude more. We compute, from the number seen through Herschel's telescope in one region of the sky, that there are a hundred millions. Imagination pierces beyond the extent of vision, beholding multitudes of unknown worlds, infinitely more in number than those which are visible to our sight; and ranges unrestrained in the boundless space of creation.

Our only difficulty with respect to the inhabitants of so many millions of planets, is the obscurity of the final causes, which it is difficult to admit when we see into what errors the greatest philosophers have fallen; for instance Fermat, Leibnitz, Maupertius, &c. in attempting to employ these final causes or metaphysical suppositions of imagined relations between effects that we see and the causes we assign them, or the ends for which we believe them to exist.

If the plurality of worlds be admitted without difficulty; if the planets are believed to be inhabited, it is because the earth is considered merely as a habitation for man, from which it is inferred that were the planets uninhabited they would be useless: but I will venture to assert that such a mode of reasoning is confined, unphilosophic, and at the same time, presumptuous. What are we in comparison of the universe? Do we know the extent, the properties, the destination, and the connexions of nature? Is our existence formed as we are, of a few frail atoms, to be considered any thing when we think of the greatness of the whole? Can we add to the perfection and grandeur of the universe? These ideas are expressed by Saussure,[44] who in speaking of a traveller to Montblanc says: 'if, during his meditations, the thought of the insignificant beings that move on the face of the earth offers itself to his mind, if he compares their duration with the grand epochs of nature, how great will be his astonishment that man, occupying so small a space, existing so short a time, can ever imagine that his being is the only end for which the universe was created.'

From these considerations d'Alembert,[45] in the *Encyclopedia,* (art. World) after examining the arguments for supposing the planets inhabited, concludes by saying: *the subject is enveloped in total obscurity.*

But Buffon affirms that wherever there is a certain degree of heat, the motion produces organized beings; we need not enquire in what way, but imagine these to be the inhabitants of the planets: if that should be the case, we may conclude it highly probable that they are inhabited, notwithstanding the preceding objections.

LALANDE.

Men and women with telescopes, gyroscopes,
a hot-air balloon and battleships.
Engraving by Étienne Claude Voysard after Claude Louis Desrais.
Reproduced by courtesy of the Wellcome Library.

PREFACE

BY BERNARD DE FONTENELLE

I FIND MYSELF NEARLY in the situation of Cicero, when he undertook to write in his own language on philosophical subjects, that, till then, had never been treated of but in Greek. He tells us that his works were said to be useless, because those who delighted in philosophy, having taken the pains to study the books written in Greek, would not afterwards think of examining his Latin ones, which were not originals; and that persons who had no taste for philosophy, would neither care for the Greek nor the Latin.

To which he answers, that exactly the contrary would happen; that the unlearned would be allured to philosophy by the facility of reading Latin works; and that the well informed, after studying the Greek authors, would be pleased to see how the subjects were handled in Latin.

Cicero might with propriety speak in this manner; his superior genius and great celebrity assured him success in this untried project, but I have not the same advantages to inspire me with confidence, in a similar undertaking. I was desirous of representing philosophy in a way that was not philosophical! I have attempted, to compose a book that shall neither be too abstruse for the gay, nor too amusing for the learned. But if what was said to Cicero should be repeated to me, I could not venture to answer as he did: possibly in attempting to find a middle way which would accommodate philosophy to every class, I have chosen one that will not be agreeable to any. It is very difficult to maintain a medium, and I think I shall never be inclined to make a second attempt of this nature.

I should warn those that have some knowledge of natural philosophy, that I do not suppose this book capable of giving them any information; it will merely afford them some amusement, by presenting in a lively manner what they have already become acquainted with by dint of study. I would also inform those who are ignorant of these subjects that it has been my design to amuse and instruct them at the same time: the former will counteract my intention if they here expect improvement and the latter if they only here seek for entertainment.

I need not say that of all philosophical subjects I have chosen that which is most calculated to excite curiosity: surely nothing ought to interest us more than to know how our own world is formed; and whether there be other worlds similar to it, and inhabited in the same way: but let no one be disquieted if unable to answer these enquiries; they who have time to spare may examine each subject; many have it not in their power.

In these Conversations I have represented a woman receiving information on things with which she was entirely unacquainted. I thought this fiction would enable me to give the subject more ornament, and would encourage the female sex in the pursuit of knowledge, by the example of a woman who though ignorant of the sciences, is capable of understanding all she is told, and arranging in her ideas the worlds and vortices.[46] Why should any woman allow the superiority of this imaginary marchioness, who only believes what she could not avoid understanding?

'Tis true, she gives some attention to the subject, but what sort of attention is requisite? Not such as will laboriously penetrate into an obscure thing, or a thing that is spoken of in an obscure manner; it is needful only to read with sufficient application to render the ideas familiar. Women may understand this system of philosophy by giving it as much attention as they would bestow on the *Princess of Cleves*,[47] in order to understand the story and see all the beauties of the work. I do not deny, that the ideas contained in this book are less familiar to the generality of females than those in the *Princess of Cleves,* but they are not more abstruse, and I am convinced that on a second perusal they would be perfectly understood.

As I did not wish to establish an imaginary system that had no foundation, I have employed true philosophical arguments, and as many of them as were necessary to establish my opinions; but fortunately the ideas connected with natural philosophy are in themselves beautiful; and whilst they satisfy the understanding, give as much pleasure as if formed only to charm, the imagination.

To such parts of my subjects as did not possess these beauties I have given extraneous ornaments; Virgil has done this in his *Georgics,* where he renders a dry subject interesting by frequent and agreeable digressions: Ovid likewise in his *Art of Love* has pursued the same plan, although the matter of his poem was far more pleasing than any thing he could add to it: he seems to think it tiresome to speak constantly of one subject—even of love. I have more need of embellishments than he, yet I have used them sparingly. I have only given such as the freedom of conversation authorised; I have only placed them in parts that I thought required them; I have inserted most of them in the commencement of the work to accustom the mind by degrees to the objects I wish to present to its attention; in short, I have derived them from my subject, or formed them as much as possible to resemble my subject.

I did not venture to give any opinions on the inhabitants of the different worlds since they must have been entirely chimerical; I have endeavoured to express all that might reasonably be imagined, and even the conjectures that are added are not without foundation. Truth and fiction are in some measure blended, but always so as to be distinguishable from each other: I do not undertake to justify such a composition; the union of philosophy and amusement is the chief aim of this work, but I know not whether I have adopted the right method.

It only remains for me now to address one class of persons; they are perhaps the moſt difficult to satisfy, not because my reasoning is inconclusive, but because they feel themselves privileged to disregard the beſt arguments: I am speaking of scrupulous people who may imagine religion is endangered by placing inhabitants any where but on the earth. I reſpeﬅ even an excessive scrupulosity when it arises from piety, nor would I willingly hurt the feelings of anyone from whom I differed: but by reﬅifying a little error of the imagination we shall find that this objeﬅion cannot effeﬅ my syſtem of giving inhabitants to an infinite number of worlds. When you are told that the moon is peopled, you immediately figure to yourself men like ourselves, and then a variety of theological difficulties occur. The poſterity of Adam cannot have colonized the moon; therefore the inhabitants of that planet are not descendants of our firſt parents; now it would be a difficult point in theology to account for the exiſtence of men who had any other anceſtor. No more need be said; every imaginable difficulty is included in this, and the expressions that would be necessary for a more full explanation are too worthy of reverence to be employed in a work containing so little of the serious as this. The objeﬅion then turns on the exiſtence of men in the moon, but it is the objeﬅors themselves who talk of men as its inhabitants. I have asserted no such thing: I say there are inhabitants, and I likewise say they may not at all resemble us. What are they then?—I have never seen them; I do not speak from acquaintance with them.

Do not consider it a subterfuge, to rid myself of the objeﬅion, when I affirm that the moon is not peopled by men; you will see that according to the idea I entertain of the endless diversity of the works of nature, it is impossible such beings as we, should be placed there. This opinion is supported throughout the book, and it is an opinion which no philosopher can deny: I think, therefore, on this ground, the following conversations will be objeﬅed to only by those who have never read them. But will this consideration suffice to deliver me from the fear of censure? No; it rather gives me cause to apprehend objections from every side.

FONTENELLE.

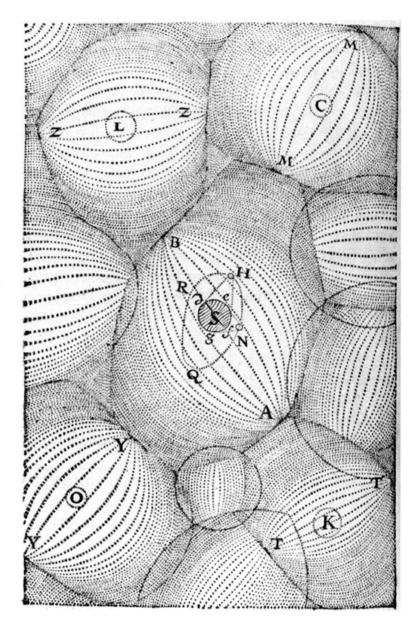

A diagram of Cartesian vortices.

Conversations on the Plurality of Worlds

TO MR. L——

You desire me, dear Sir, to give you a particular account of the manner in which my time has been spent whilst at the Marchioness of G——'s* in the country. To obey your injunctions strictly, I shall be obliged to fill a volume, and what is still more formidable, a volume of philosophy.

You expect to be entertained with a history of splendid feasts, hunting, and card-parties; and you will hear of nothing but planets, worlds, and vortices:† for the discussion of these latter subjects formed our principal amusement. Fortunately *you* are a philosopher, therefore I have the less reason to dread raillery from such a quarter; on the reverse, I may even hope for your congratulations, on having rendered the Marchioness sensible to the charms of philosophy; we could not have made a more valuable acquisition; for youth and beauty, in every cause, holds such power, that if Wisdom herself were desirous of being welcomed by mortals, and would assume the form of this lovely woman,

---

* The lady here mentioned was Madame de la Mesangire [Mésangère] of Rouen. She was a beautiful brunette; but in compliance with her desire to be concealed, the author has spoken of her in the following pages, as having a fair complexion. The park belonging to her residence, is described in the 'First Evening'.

† The Vortices of Descartes occupied the attention of the learned for nearly a century; but this hypothesis was superseded by the discovery of the laws of attraction. Although Newton's famous book on principles was published in 1687, Fontenelle always retained his educational prejudice in favour of the Vortices. A few years before his death he consulted me on a little work he had some time since composed on the subject. I endeavored to dissuade him from making it public; but Falconet was afterwards weak enough to do so. The book is entitled, *Theory of the Cartesian Vortices, with Reflections on Attraction.* The author's name was never affixed to the work.

surely with such an exterior, and such fascinating eloquence, she could not fail to attract every heart.

Notwithstanding all this, you must not expect to be transported with admiration, whilst I repeat the conversations I have held with her ladyship; my genius should be equal with hers, to relate what she said, in her own delightful manner. Conscious of inability, I must relinquish the attempt, and leave you to discern through the recital, that rapidity of apprehension, which characterizes the mind of the Marchioness. From the wonderful quickness with which she comprehends the most abstruse subjects, I consider her already learned: at least I may be allowed to say, that after a little study, she might attain the heights of science; when many who spend their lives amid the dull disputes of vast libraries remain for ever in the deepest ignorance.

Before I recount our various conversations, perhaps you may expect some description of their scene; some picture of the romantic country, under whose shades the Marchioness is enjoying the autumn. If so, you will be disappointed: so many people have exercised their talents on this gay species of writing, that I shall dispense with the ceremony; and merely say, that on my arrival I had the pleasure of finding myself the only visitor.

The two first days were passed in relating the news of Paris, which I had just quitted. When that subject was exhausted, an evening walk in the park suggested the discussion of those learned topics, the commencement of which you will find in the next page.

Ƒirst Evening[*]

The Earth is a Planet which Turns on its Axis, and Goes Round the Sun

A FTER SUPPER WE WENT to take a walk in the park. We felt the fragrant breeze of evening peculiarly delightful, as the heat had been intense during the day: the silvery rays of the moon, gleaming through the foliage, formed an agreeable contrast with the darkened shadows of the landscape. Not a cloud intercepted or veiled the smallest star. Every orb appeared a mass of pure gold, rendered more brilliant by the rich blue sky. The beauty of the scenery produced a gentle reverie, from which, had not the Marchioness been with me, I should not have been easily roused; but in the company of so interesting a woman I could not long abandon myself to the influence of the moon and stars. Do you not think, said I, addressing myself to her, that the charms of a fine night greatly exceed those of the day? Yes, she replied, the splendor of day resembles a fair and dazzling beauty, but the milder radiance of night may be compared to a woman of less brilliancy of complexion, and more sweetness of expression. You are very generous, resumed I, in giving the preference to the brunette, while you are so fair. It is, however, true, that an unclouded sun is the most glorious object in nature; and it is equally true, that the heroines of romance, the most beautiful objects imagination can depict, have almost invariably been represented with fair complexions. Beauty, answered my companion, is nothing, unless it interests our feelings. You will not deny that the finest day never had the power of

[*] This first book has been translated into a variety of languages; it is the best eclogue that has been composed in the last fifty years: the descriptions and imagery it contains are perfectly suited to the style of pastoral poetry; indeed many of the images would not have disgraced the pen of a Virgil.—Dubos. *Reflections on Poetry and Painting.*

inspiring so delightful a reverie as you were falling into juſt now in contemplating the loveliness of the evening. You are right, said I, but the lovelieſt of the night I ever beheld, with all its shadowy beauty, would fail to give me such enchanting sensations as the contemplation of the fair face of the Marchioness de G———. I should not be satisfied with your compliment, she replied, did I even believe you sincere, since the brightness of day, with which we have been comparing fair women, has so little influence on your heart. Why do lovers, who undoubtedly can judge of what is moſt touching, address all their poetic effusions to the night? To the ear of day they neither confide their transports nor their sorrows—why is it so entirely excluded from their confidence? Probably, I answered, because it is not calculated to inspire that delicious sentiment, at once impassioned and melancholy, which we feel in the ſtillness of night, whilſt all nature seems to repose. The ſtars appear to move with more silent progress than the sun: every objeƈt that decorates the heavens is soft, and attraƈtive to the eye; in short, we resign ourselves more easily to reverie, because we feel as if no other being was at that time enjoying the pensive pleasure that expands our soul. Perhaps, too, the uniformity of day, in which the sky presents no other objeƈt than the sun, is less favorable to the wild and pleasing illusions of fancy, than the view of innumerable ſtars, scattered with sportive irregularity over the boundless space. I have always felt what you describe, said she, I love to see the ſtars, and am almoſt inclined to reproach the sun for hiding them. Ah! cried I, I cannot forgive him for concealing so many worlds from my sight! Worlds! she exclaimed, turning to me with surprize, what do you mean? Forgive me, said I, you touched the wildeſt chord of my imagination.——I forget myself in a romantic idea. And what is this romantic idea? enquired the Marchioness. Ah! replied I, I am half ashamed of owning it: I have taken it into my head that every ſtar may be a world. I would not positively assert the truth of my opinion, but I believe it because it affords me pleasure; it has possessed my mind with irresistible force; and I consider pleasure a needful accessary to truth. Well, said she, since your whim is such a pleasant one, make me a partaker of it; I'll believe any thing you chuse about the stars, provided it contributes to my happiness. Ah! madam, I replied, 'tis not such an enjoyment as you would find in seeing one of Moliere's comedies: it is an idea which can only give delight to the understanding. What! exclaimed she, do you think I am not susceptible of pleasures which depend only on reason? I will convince you of your mistake. Teach me your system. No, answered I, I will not subject myself to the reproach of having talked of philosophy, in such an enchanting walk as this, to the most interesting woman of my acquaintance. No, seek for pedants elsewhere.

For a long while I, attempted, in vain, to excuse myself; I was at last obliged to yield. I insisted, however, for my reputation's sake, on a promise of secrecy. Every objection being removed, I wished to begin the subject, but found the commencement extremely difficult; for, with a person who was ignorant of natural philosophy, it was necessary to converse in a very circuitous manner, to prove that the earth was a planet, the other planets similar to the earth, and all the stars so many suns, which enlightened a number of worlds. I once more assured her it would be much better to talk on such trifles as other people in our situation would amuse themselves with. In the end, however, to give her a general idea of philosophy, I pursued the following plan.

All philosophy, said I, is founded on two things; an inquisitive mind, and a defective sight; for if your eyes could discern every thing to perfection, you would easily perceive whether each star is a sun, giving light to a number of worlds; on the other hand, had you less curiosity, you would hardly take the trouble to inform yourself about the matter, and consequently remain in equal ignorance; but the difficulty consists in our wanting to become acquainted with more than we see: besides, it is out of our power to understand much of what is even within the reach of our sight, because objects appear to us very different from what they are. Thus philosophers pass their lives in disbelieving what they see, and endeavouring to conjecture what is concealed from them; such a state of mind is not very enviable.

In thinking on this subject, nature always appears to me in the same point of view as theatrical representations. In the situation you occupy at the opera you do not see the whole of the arrangements: the machinery and decorations are so disposed as to produce an agreeable effect at a distance, and at the same time the weights and wheels are hidden, by which every motion is effected. You behold all that is passing, without concerning yourself about the causes; and so perhaps do all the other spectators, unless among the number some obscure student of mechanics is puzzling himself to account for extraordinary motion which he cannot understand. You see the case of this mechanical genius resembles that of the philosopher studying the structure of the universe. What, however, augments the difficulty with respect to philosophers is, that nature so entirely conceals from us the means by which her scenery is produced, that for a long time we were unable to discover the causes of her most simple movements. Figure to yourself, as spectators of an opera, the Pythagorases, the Platos, the Aristotles; and these men whose names are now so celebrated. Let us suppose them viewing the flight of Phaeton, rising on the wind; ignorant at the same time of the construction of the theatre, and the cords by which the figure is put in motion. One to explain the phenomenon, says, *it is some hidden virtue in Phaeton which causes him to rise;* another replies, *Phaeton is composed of certain numbers which produce its elevation.* A third says, *Phaeton has a love for the top of the stage; he is uneasy at any other part.* The fourth thinks, *it is not essential to the nature of Phaeton to rise in the air, but he prefers flying up to leaving a vacuum at the top of the stage.* Such were the ridiculous notions of the ancient philosophers, which, to my astonishment, have not ruined the reputation of antiquity. After all, Descartes and some other moderns appear: they tell you that *Phaeton rises in consequence of being drawn by cords, fastened to a descending weight, which is heavier than himself.* It is no longer believed that a body can have motion, unless acted upon by another body; that it can rise and descend without a counterbalancing weight; thus, whoever examines the mechanism of nature, is only going behind the scenes of a theatre. If that be the case, answered the Marchioness, philosophy is a very mechanical affair! So much so, I replied, that I am afraid it will fall into disrepute. In short, the universe is but a a watch on a larger scale; all its motions depending on determined laws, and the mutual relation of its parts. Confess the truth, have you not hitherto entertained, a more exalted idea of the works of nature? Have you not considered them with more veneration than they deserve? I have known some people esteem them less as their knowledge increased. For my part, said she, I contemplate the universe with more awful delight now I find that such wonderful order is produced by principles so simple.

I know not, rejoined I, how you have acquired such rational ideas, for, to say the truth, they are not very common. The generality are affected only by the obscure and marvellous. They admire nature merely because they consider it a sort of magic: something too occult for the understanding to reach: to them a thing appears contemptible as soon as they find the possibility of explaining its nature: but you, madam, can reason so clearly, that I have only to draw aside the veil, and present the world to your inspection.

What we behold at the greatest distance from our earth is the azure heaven, that immense arch to which the stars seem firmly to adhere. They are called fixed, because they appear to have no other motion than that of their sky, carrying them from east to west. Between the earth and the remote firmament are suspended, at various distances, the Sun, Moon, and other five stars, de-

nominated planets; Mercury, Venus, Mars, Jupiter, and Saturn.[*]
These planets not being stationary at one point in the heavens, but
having unequal motions, vary with respect to their relative situations; the fixed stars, on the contrary, always bear the same local
relation to each other. The chariot,[48] for instance, that you may
distinguish, formed of those seven stars, has always had that con-
figuration, and is likely to retain it; but the moon sometimes approaches nearer to the sun; sometimes retreats farther from it; the
same is observed of the other planets. Such were the observations
made by the Chaldean[49] shepherds, whose continual leisure enabled them to give so much attention to the heavenly bodies, as to
form the rudiments of astronomy, for we learn that science took
its rise in Chaldea,[†] as geometry was first studied in Egypt, where
the inundations of the Nile destroyed the boundaries of different
possessions, made the inhabitants desirous of exact measures, by
which they could again separate their own lands from those of
their neighbours. Thus astronomy is the offspring of idleness, geometry of interest; and if we enquire into the origin of poetry, we
shall probably find that she is the daughter of love.

I am glad, said the Marchioness, you have given me this geneal-
ogy of the sciences; astronomy is the only one that will suit me:
geometry, according to your account of it, requires a more selfish
heart than mine; and I have not susceptibility enough to attempt
poetry with success; I have, however, as much leisure as can be
needful for the study of astronomy; it is another favourable circumstance that we are in the country, leading a pastoral life. Do
not mistake, Madam, answered I, talking of planets and fixed stars
is not all that constitutes a pastoral life. Was the conversation of
the shepherds, in the golden age, confined to astronomy? Ah, said
she, but it would be dangerous to conform one's mode of life to
theirs. No, that of the other shepherds you mention appears pref-

[*]   In 1781, M. Herschel discovered a sixth. *Astronomy* by Lalande, third edition, 1792,
Vol. 1, Art. 116.

[†]   Perhaps in Ethiopia. *Astronomy*, Art. 260.

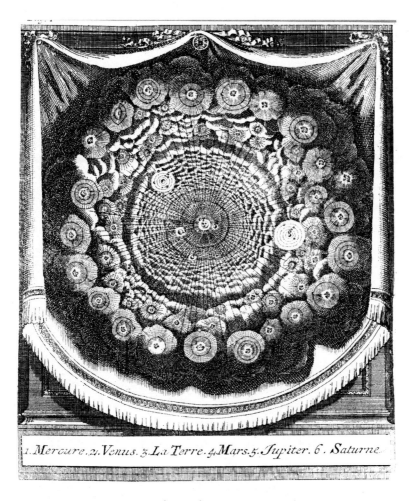

1. *Mercure*. 2. *Venus*. 3. *La Terre*. 4. *Mars*. 5. *Jupiter*. 6. *Saturne*.

Mercury, Venus, the Earth, Mars, Jupiter and Saturn.
Engraving from *Entretiens sur la Pluralité des Mondes*, Amsterdam, 1719.
Uranus was not discovered by William Herschel until 1781.

erable to me; therefore, let us converse, if you please, in Chaldean
ſtyle. After this disposition of the ſtars was remarked, what followed? The next thing, I replied, was to imagine the arrangement
of the different parts of the universe; that is what the learned call
making a syſtem. But before I explain to you the firſt of these systems, give me leave to premise that we are all naturally disposed
to the same sort of madness as a certain Athenian of whom you
have heard, who had taken it in his head that every vessel which
went into the port of Pyreum[50] belonged to him.[51] We chuse to
believe that every thing in creation is deſtined to our service; and
when we enquire of some philosophers the use of such a prodigious number of fixed ſtars, of which a smaller proportion would
have been sufficient for all the offices they appear to perform; they
coolly answer, they were made to gratify our sight. On this selfish principle it was for a long time supposed that the earth was
motionless in the midſt of the universe, whilſt all the heavenly
bodies were created for the sole purpose of journeying round,
and diſtributing their light to her. Next to the earth they placed
the moon, after the moon, Mercury, then Venus, the Sun, Mars,
Jupiter, Saturn; beyond all these the firmament of fixed ſtars. It
was imagined that the earth was ſtationed exaĉtly in the middle of
the circles described by these planets, which extended in proportion to their diſtance from the earth, and that, consequently, the
moſt remote planets required a longer time to perform their revolutions, which certainly is true. But, interrupted the Marchioness,
I can't see why you would disapprove such an arrangement of the
universe; it appears to me sufficiently commodious and intelligi-
ble; I really feel quite satisfied with it. I have taken pains, answered
I, to represent this syſtem in the moſt favourable point of view;
if I were to explain it exaĉtly as it was, conceived by Ptolomy, the
author of it, and his disciples, you would be quite shocked. As
the motions of the planets are irregular, being sometimes quicker,
sometimes slower; going sometimes in one direĉtion, sometimes

in another; now nearer to the earth, then at a greater distance from it; the ancients figured to themselves an endless number of circles intersecting each other, by which they endeavoured to understand the great variety of movements. The confusion, however, caused by such an infinity of circles was so perplexing, that, at the time no better system was known, one of the kings of Castile,* a profound mathematician, was daring enough to say, that if the Supreme Being had consulted him when he created the world, he would have given him some good advice. We are filled with horror at the impiety of this expression, but it serves to shew us how absurd must have been the hypothesis which could prompt it. The advice this man wished to have given, undoubtedly regarded the suppression of so many circles, which did but prevent the planetary motions from being understood. Probably he would likewise have expunged from the system two or three superfluous firmaments, supposed to be above the fixed stars. The philosophers, to explain some particular motion of the heavenly bodies, placed, beyond the heaven that bounds our view, a sky of crystal, which communicated this motion to the lower sky. Was a new movement discovered? They had nothing to do but to form a second crystal firmament. In short, skies of crystal were made without any trouble. Why did they always chuse crystal? enquired the Marchioness; would nothing else have answered the purpose as well? No, answered I, it was necessary to have a substance, at once transparent and solid, for it was Aristotle's opinion that solidity was essential to the dignity of their nature, and as this was believed by a great man, nobody thought of doubting it. But since that time comets have been seen, which, being higher than was formerly imagined, must have broken all the crystal of these skies, in passing through them, and by that mean, thrown the universe into confusion; it was therefore found necessary to change the matter of which these firmaments were composed, into a fluid, such as air.

*   Alphonsus, king of Castile, died 1284.

It is now discovered with certainty, by the researches of later ages, that Venus and Mercury turn round the Sun, and not round the earth, on this subject the ancient system is absolutely exploded. I will now acquaint you with another which provides for every difficulty, one that does not require any amendments of the king of Castile, for its simplicity is so charming that one cannot refuse to believe it. Yours, interrupted the Marchioness, seems a sort of bargaining philosophy; whoever offers a system that is effected at the least expence, has the preference. 'Tis true, said I; we have no other chance of understanding the plan by which the operations of nature are carried on. Nature is a wonderful economist; if a work is to be effected, and two ways are practicable, we may be sure she will adopt that which costs her the least, however trifling the difference. This economy is notwithtanding in every respect consistent with the surprising magnificence which appears in all her productions. Magnificence is employed in the design, and frugality in the execution of it. Nothing should excite our admiration so much as a stupendous project effected by simple means: but we are apt to cherish ideas of a very different kind. We place the frugality in the designs of nature, and her grandeur in the execution. We imagine her forming a contracted plan, and executing it with ten times the labour that is requisite: what can be so ridiculous? I hope, she replied, that the system you are going to explain will strictly imitate nature; the simplicity you so admire will spare me a great deal of trouble in comprehending your instructions. Your hope will be realized, said I, we have now no useless incumbrances. At the appearance of a certain German named Copernicus,[*] astronomy became simplified; he destroyed all the unnecessary circles, and crushes to pieces the crystaline firmament.[†] Animated with philosophic enthusiasm, he dislodged the earth from the central situation which had been assigned it, and in its room placed the sun,

[*]   He was born in 1472, at Thorn, in Prussia Royal; he died in 1543.

[†]   Several of the ancients were of opinion that we should admit the motion of the earth. *Astron.* Art. 1075.

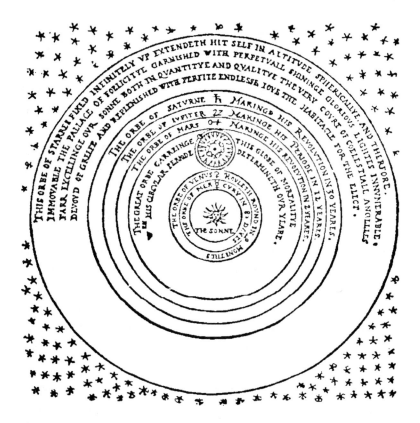

Copernicus's heliocentric universe.

who was more worthy of such a mark of diſtinction. The planets were no longer supposed to perform their revolutions round the earth, and enclose it in the centre of their orbits. If they afford us light it is as it were by chance, and in consequence of passing us in their course. They all turn round the sun; the earth itself not excepted; and as a punishment for the indolent repose it had been thought to enjoy, Copernicus made it take an ample share of the general activity: in short of all these celeſtial attendants, appointed for the service of our little globe, the moon alone is left to move round it. Stop a moment, said the Marchioness, your

imagination is so elevated with your subject, you have explained
it in such pompous language, that I believe I have scarcely understood you. The sun, you say, is immoveable in the centre of the
universe; which of the planets is next in succession? 'Tis Mercury,
I replied. Mercury goes regularly round the Sun in nearly a circular orbit, of which that luminary is the central point. Next to
Mercury is Venus, which turns in the same manner round the Sun.
Afterwards comes the Earth, and being higher than Mercury and
Venus, describes a larger circle round the Sun than either of those
planets. Then follow Mars, Jupiter and Saturn, in the order I have
named them; thus you see the circle of Saturn must be the most
extensive of all; it likewise requires a longer time than the other
planets to perform its revolution. But, exclaimed she, you have
forgotten the moon. I shall recollect it presently, said I; the moon
never abandons the earth, but is constantly going round it; but as
the earth is continually moving onwards in a circle round the sun,
the moon at once follows its motion, and revolves round it; this attendant planet, therefore only goes round the sun in consequence
of invariably continuing near to the earth.

I understand you, said she, and I love the moon for remaining
attached to us when we were forsaken by all the other planets.
Confess that if your German could have alienated that too, he
would have done it without regret; for one may see, in every
part of his hypothesis, that he was but little inclined to favour the
earth. He did well, I replied, to humble the vanity of men who
chuse to claim the best situation in the universe; 'tis with pleasure I consider the world mixing in the croud of planets. Pshaw!
cried the Marchioness; do you think vanity can have any thing to
do with a system of astronomy? Do you suppose I feel humbler
for knowing that the earth goes round the sun? I assure you I esteem myself just as highly as I did before. Certainly, madam, I
answered, men would be less concerned about the rank they held
in the universe than that they enjoyed amongst their associates;

neither would the disputes of two planets, for precedence, be so important in their judgment as those of two ambassadors. Nevertheless, the same disposition which induces a man of the world to aspire after the moſt honourable place in a room, will make a philosopher desirous of placing the globe on which he lives in the moſt diſtinguished situation in the universe. He likes to consider every thing made for his use; he encourages, without, being aware of his vanity, so flattering an opinion, and his heart becomes deeply engaged about an affair of mere speculation. Upon my word, she exclaimed, you are calumniating human nature—how happened it that the opinions of Copernicus were received, since they are so very humiliating? Copernicus, answered I, was himself very doubtful of the reception they would meet with; it was a long while before he could resolve to publish his syſtem; at laſt, however, he yielded to the entreaty of several diſtinguished charaċters; but what do you think was the consequence?—on the day they took him the firſt printed copy of his book, he died: so he made use of the shorteſt way to escape from all the contradiċtions he had been anticipating.

Liſten to me, said the Marchioness: let us be juſt to every body: it certainly is difficult to imagine we turn round the sun, for we never change places; we rise in the morning where we went to reſt at night—I see from your look you are going to tell me, that as the earth moves altogether—Assuredly, said I, it is the same thing as going to sleep in a boat which was sailing down a river; on waking, you would find yourself in the same boat, and in the same part of the boat. Yes, replied she, but here is a difference; when I awoke I should find an alteration in the shore, and that would prove that the boat had changed its place: it is not so with respeċt to the earth, I there find every thing as I left it. No, no, madam, you may observe an alteration in the shore, as you do in the boat: you will recolleċt that beyond the planets are the fixed ſtars, these are what we muſt consider the objeċts on the shore. I am on the earth;

the earth describes a large circle round the sun; when I look to the middle of the circle I find the sun, and were not its brightness so dazzling as to render the ſtars invisible, I should conſtantly see, by looking beyond the centre, some of the fixed ſtars opposite to the sun; by viewing them however at night, I can easily determine how they were situated in the day, which renders my observations equally accurate. If the earth remained in the same place I should always find the same fixed ſtars answering to the situation of the sun, but as the earth moves on in her orbit, I necessarily see other fixed ſtars at the point which I had before examined. Such is our shore, which is every day varying; and as the earth goes round the sun in a year, I observe the sun, during that space of time, successively answering to different fixed ſtars which compose a circle; this circle is called the Zodiac; shall I shew it you by tracing a figure on the earth? No, said she, I can, dispense with that; it would give my park too learned an appearance. I think I have heard that a philosopher, shipwrecked on an island with which he was unacquainted, cried out to his companions on perceiving some lines and circles drawn on the sand: *courage, my friends! the island is inhabited; here are footſteps of men.* You muſt consider that such *footſteps* ought not to be seen here.

It would certainly, I replied, be more in charaĉter, to trace only the footſteps of lovers; that is to say, your name, engraved by your adorers on the bark of every tree. No more of adorers, cried she; let us talk of the sun. I underſtand perfeĉtly why we imagined it performing the revolution which is merely effeĉted by our own motion, but this circle requires a year; how, then, does the sun appear to go round us every day? You undoubtedly know, I replied, that if a ball were rolled along this walk it would have two sorts of motion; it would go towards the end of the walk, and at the same time turn several times on its own axis, so that the side of the ball which is at firſt uppermoſt will presently descend, and the other, of course, rise to the top. This is the case with the earth:

whilst she is proceeding, through the year, in her orbit round the sun, she turns on her own axis every twenty-four hours: each part, therefore, of the earth loses and recovers sight of the sun during that time. When in the morning we turn towards the sun, it seems to rise; and when, by continued rotation, we again are more distant from it, it appears to descend. That's curious enough, said she; the earth undertakes every thing, and the sun does nothing at all: and when the moon, the other planets, and the fixed stars appear to pass over us in four-and-twenty hours, it is merely imagination. Exactly so, I replied, and produced by the same cause. The planets perform their revolutions round the sun in unequal periods of time, in consequence of their different distances from it; and one which we see today answers to a certain point in the zodiac, or circle of fixed stars, will tomorrow at the same hour answer to some other point: this is the effect both of the progress which the planet has made in its orbit and of that which we have made in ours. We move onward; the other planets do the same, but we do not continue all in a line: this occasions us to see them in such different situations, and renders their course apparently irregular. You now understand that such irregularity depends only on the different points from which we view them, and that, in reality all their movements are directed by the most exact order. It may be so, answered the Marchioness; but I should be glad if their order did not cost the earth so much: you seem to have very little consideration for it, and to require an astonishing agility in so large a body. But, said I, do you think it more reasonable for the sun, and all the other heavenly bodies, which are extremely large, to perform in four-and-twenty hours an immense journey round the earth? That the fixed stars, being in the largest circle, should travel, in the course of a day more than twenty-seven thousand six hundred and sixty times two hundred millions of leagues?[*][52] All

* According to later calculations, it would be a thousand millions of times a million of leagues; but a person who did not believe the motion of the earth would have no occassion to admit this prodigious distance.

this muſt be if the earth does not turn on her axis every twenty-four hours. Surely there is much more rationality in supposing our globe to turn, which would give to each part but a journey of nine thousand leagues. Consider what a trifle are nine thousand in comparison of the terrific number I have juſt mentioned.

Oh! replied the Marchioness; the sun and ſtars are made of fire, a swift motion is nothing to them, but the earth does not seem *formed for motion*. And should you think, I answered, if experience had not proved the faɛt, that a large ship, carrying a hundred and fifty pieces of cannon, three thousand men, and a heavy freight of merchandize, could be formed for motion? Yet a gentle breeze is sufficient to move it forwards, because the water, being liquid, and easily divided, makes a very slight resiſtence to the progress of the ship, or if it be in the middle of a river, it follows without difficulty the current of the water, since there is then no impediment. In like manner the earth, notwithſtanding its weight is with facility carried through the sky which is infinitely more fluid than water, and which fills the immense space occupied by all the planets. To what could the earth be faſtened so ſtrongly as to resiſt the current of this celeſtial fluid and remain motionless? we might as well imagine a little wooden ball could resiſt the tide of a river.

But, said she, how can a body so ponderous as the earth be suspended in your celeſtial fluid, which from its great fluidity muſt be extremely light?

It does not follow, I replied, that a subſtance muſt be extremely light because it is fluid. What do you think of the great ship we have been talking of, which, with all its lading, is lighter than the water that supports it? I won't have any more to say to you, answered she, half angrily, if you mention your ship again.* But tell me, is it not dangerous to inhabit such a whirligig as you represent the earth? If you are afraid, said I, let us have the world supported by four

---

* The Marchioness was in the right not to liſten to such an answer. It is absurd to pretend that the ætherial fluid, so light and rare, can be capable of bearing along those enormous masses, the planets.

elephants, as the Indians do. Well! cried she, here is a new syſtem. I like those people for providing such good foundations for the earth to reſt on, whilſt we Copernicans are imprudent enough to swim at random in this celeſtial fluid. I dare say if the Indians knew there was the leaſt danger of the earth's being moved, they would double the number of elephants.

It would be worthwhile, I replied, laughing at the thought, we muſt not be sparing of elephants if they can enable us to sleep in peace; if you would find them serviceable tonight we'll put as many as you please, and then remove them one or two at a time as you find your courage return. No, said she, I don't think there is any need for them, and to speak seriously, I feel courageous enough to let the world turn round. And I can venture to prediſt answered I, that in a little while its turning will give you pleasure; will even inspire the moſt delightful ideas. I sometimes imagine myself raised above the surface of the earth, and remaining motionless whilſt its daily rotation continues. All the different inhabitants pass in review; some fair, some copper-coloured, some black. Now I see heads covered with hats; then with turbans; some shaven, others with flowing hair. As the towns pass before me, I observe some have ſteeples, some long spires with crosses on them, others are ornamented with towers of porcelaine. Then I behold large countries with no other buildings than little huts: afterwards, immense seas; then frightful deserts; and in short, all the boundless variety which is to be found on the face of the earth.

Really, she replied, it would be worth while to devote fourand-twenty hours to such a sight. If I underſtand you, when we move, other countries with their inhabitants pass into the situation we are leaving, and so on, till in four-and-twenty hours we again arrive at the same place.

Copernicus himself, said I, could not have comprehended it more clearly. The firſt that would succeed us* would be the English: we should probably find them arguing on some political topic with less gaiety than we are discussing our philosophy. When we had dismissed them we should discover a vaſt sea,† on which perhaps would be some vessel less at her ease than we. Then come the Iroquois, eating one of their prisoners of war,[53] who does not even utter a groan though ſtill alive when they begin to devour him.‡ After them the women of Jesso, who employ all their time in preparing viſtuals for their husbands, and painting their lips and eye-brows blue to appear handsome in the eyes of the moſt disguſting men in the world. Then the Tartars, very devoutly going on a pilgrimage to their high prieſt, who dwells in an obscure recess, enlightened only by lamps, the rays of which direſt these votaries to the objeſt of their adoration. Afterwards the beautiful Circassians who make no ceremony of granting all their favours to the firſt that solicits them, except what they believe the essential prerogative of their husbands. Then the inhabitants of Little Tartary, who go to ſteal women for the Turks and Persians. Laſt of all our countrymen, whom we should find entertaining each other with the vagaries of their imagination.

It is amusing enough, said the Marchioness, to fancy one-self in a situation to see all these things: but if I were taking the view I should wish for the power of haſtening or retarding the earth's motion, according to the feelings with which each ob-jeſt inspired me: I assure you I should soon push on those that argued on politics, and the others that devoured their enemies; but there are some of the people you have been speaking of that would excite my curiosity; the handsome Circassians, for inſtance, whose cuſtoms are so peculiar. But a serious difficulty occurs to

---

*   To speak more properly they would be one hundred leagues northward.

†   The Atlantic

‡   We should next see the Pacific Ocean.

me respecting your system. If the earth turns, we every moment change our atmosphere, and respire that of a new climate. By no means, madam, I replied; the air which surrounds the earth rises only to a certain height, twenty leagues perhaps at farthest;* this atmosphere always turns with us. You have doubtless observed the sort of shell in which a silk-worm imprisons itself, and which it forms with such astonishing art. It is composed of silk closely woven, but covered with a light down. Thus it is with regard to the earth, it is a solid body covered with an atmosphere extending to a certain height, which adheres to, and moves with it, as the down does with the firmer, substance beneath it. Above our atmosphere is the celestial matter, incomparably more pure, subtle and active than air.

You represent the earth in a very contemptible light said the Marchioness. Nevertheless on this silk-worm's shell we find stupendous works, furious wars and universal agitation. Yes, answered I, and while all this is going on, nature, who does not concern herself about such frivolous things, carries us all along; with an uninterrupted motion, and amuses herself with the little ball.

It appears very ridiculous, replied she, to give way to so much anxiety, whilst one is living on a thing that is perpetually turning; but unfortunately we are not assured that it does turn, for, to tell you the truth, all the precautions you take to convince me that we should not feel its motion are unsatisfactory. How could it avoid giving some indication by which we should be sensible of it?† The most common and natural motions, said I, are the least felt: this observation is true even in a moral sense: the motions of self-love are so frequent in our minds, that for the most part we are not sensible of them, but believe ourselves actuated by other principles. Ah! you are beginning to moralize, said she; moralizing and ex-

---

* Even at two leagues it is no longer discernible.

† There are several; one of them is the aberration of the stars. *Astron*. Book xvii.

plaining natural philosophy are a little different; I fancy you grow tired of your subject: let us return home, we have had enough, for the first lecture. Tomorrow we will come here again—you with your systems, and I with my ignorance.

In our way to the house, to give her a complete history of systems, I told her that a third had been invented by Ticho Brahe, who, from a determination to keep the earth in a state of rest, placed her in the centre of the universe, making the sun revolve round her, and the planets round the sun; for in consequence of some discoveries which had been lately made, he found it impossible to make the planets go round the earth. The Marchioness, with her usual discernment, concluded that it was mere whim to exempt the earth from moving round the sun, when so many other larger bodies were allowed to perform the revolution; that the sun was rendered more unfit for turning round the earth when incumbered with the other planets, and in short that this system was only calculated to support a prejudice in favour of the earth's immobility,* without offering anything to convince the judgment; we therefore resolved to retain that of Copernicus, which is more uniform and pleasing, and at the same time unmixed with prejudice. In fact its simplicity convinces, whilst its boldness excites admiration.

* The ridiculous system of Tlcho was invented out of respect to holy scriptures, not considering that their object was more important than the refutation of popular errors in natural philosophy.

Tycho Brahe (Tyge Ottesen Brahe), the Danish astronomer.

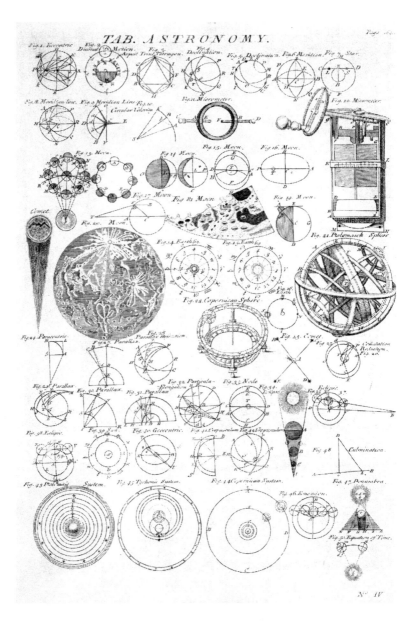

Part of the Table of Astronomy from the *Cyclopaedia,*
or Universal Dictionary of Arts and Sciences,
London, 1728.

Second Evening

The Moon is an Habitable Globe

THE NEXT MORNING, as soon as the Marchioness was awake, I sent to enquire how she did, and whether she had been able to sleep whilst the globe was turning? I received for answer, that she already felt quite accustomed to the motion; and had slept as undisturbedly as Copernicus himself. Soon afterwards, some company came to spend the whole day with her; a tiresome custom which is always observed in the country; yet long as the visit was, we considered it a great kindness in the guests, not to prolong it to the next day; which I find is a common practice in this part of the world; however, as they had the civility to leave us, the Marchioness and I had the evening to ourselves. We immediately went to the park and resumed our astronomical conversation. She understood so perfectly all I had said on the former evening, that she disdained to hear any repetition of the subject, and desired me to enter on a new one.—Well then, said I, since the sun, which we conclude is immoveable, can no longer be considered a planet; and the earth is proved to be one, and to move round the sun; you will be the less surprised to hear, that the moon is a world like ours; and to all appearance, inhabited. —I never heard speak of peopling the moon; she replied, but as a ridiculous, visionary hypothesis. It may be so, answered I; I only adopt the interest of any party, in these cases, as people do in civil wars; in which the uncertainty of the event induces them to hold correspondence with opposite sides, and even, when possible, with their enemies. For my part, though I believe the moon is inhabited, I can be very civil to anyone that disbelieves it; and

I always retain the power of going over to their side without disgracing myself, if I found they had the advantage: but in the present state of the question I have the following reasons for thinking the moon is inhabited.

Let us suppose that no communication had ever been carried on between Paris and St. Dennis; and that a Parisian[54] who had never gone out of his own city should stand on one of the towers of Notre-Dame, and at that distance view St. Dennis: were he asked if he believed that St. Dennis was inhabited like Paris, he would without hesitation answer, No; I see inhabitants in Paris, but I can discover none at St. Dennis, nor did I ever hear of any being there. Somebody standing by, might answer, that we certainly cannot see them from the tower of Notre-Dame, but that is, because we are at too great a distance; that from all we can discern of St. Dennis it is very much like Paris; that it has steeples, houses, walls; and therefore is very probably inhabited. All this makes no impression on our citizen; he insists upon it that St. Dennis is uninhabited because he does not see any body in it. The moon is our St. Dennis, and each of us is this Parisian who has never left the city in which he resides.

Oh! you wrong us, interrupted the Marchioness; we are not so stupid as your citizen; when he sees that St. Dennis is constructed exactly on the same plan as Paris, he must be out of his senses not to believe it inhabited: but the moon is very different from the earth. Be cautious, madam, said I; if the moon's resemblance to the earth prove it habitable, I shall force you to believe that it is inhabited. I confess, answered she, that if you can shew me the similarity, I cannot pretend to deny its being inhabited, and I see so much confidence in your looks that I am afraid you will be triumphant. The two different motions of the earth, which I never before knew anything about, make me fearful of hastily rejecting any other opinion; but still, can it be possible that the earth is luminous like the moon?—that you know is essential to their similarity. Indeed, madam, I replied, the luminous quality of planets depends on less than you imagine. The sun alone is, in his *nature*, luminous; but the planets only reflect the light they receive from him. He enlightens the moon; the moon reflects his rays on the earth, and the earth is undoubtedly in the same manner a source of light to the moon; it is no farther from us to the moon, than from the moon to us.

But, enquired the Marchioness, is the earth equally capable of reflecting the sun's light? I see, answered I, you have an invincible partiality for the moon. Light is composed of globules which rebound from a solid substance, but pass through any thing in which they find interstices, such as air or glass: the moon, therefore, gives us light in consequence of being a hard, solid body, which sends back these globules. I suppose you will not dispute the hardness and solidity of the earth. See then the effects of an advantageous situation—because the moon is at a distance we only view her as a luminous body instead of a large mass of matter similar to the earth. Our globe, on the contrary, from having the ill luck to be more closely inspected, appears only a mass of dark soil, fit for nothing but to produce food for animals; we do not perceive the splendour of her light, because we cannot remove to a distance from her. So it is, answered the Marchioness, with the different ranks of society: we are dazzled with the grandeur of situation, superior to our own, without considering how much every condition of human life resembles all the rest.

'Tis precisely the same thing, I replied; we take upon us to decide on every thing, but we are never in a proper place for making our observations. We would form an opinion of ourselves, and we are too near; we would judge of others; they are too distant from our view. We should be placed between the earth and the moon to form a just comparison; as a spectator, not an inhabitant of the world. I shall be inconsolable for the injustice we do our world, said she, and the partial regard we have for the moon, unless you can assure me that the inhabitants of that planet, are as ignorant of *their* advantages, and consider our globe a luminous body, without knowing that from their own we derive so much light. I can make you easy on that head, answered I; we are certainly a luminary to them: they do not, it is true, see us describe a circle round them,* but that does not signify. The reason of our appearing to remain in the same place is this:—the side of the moon which was turned towards us at the creation, has always continued so; we always observe the same eyes, mouth, and other features of the face which, by the help of imagination, we have contrived out of the spots on her surface.† If the other half were presented to us, we should see spots arranged in a different form: this does not arise from the moon's not turning on her axis; she turns in the same time that is employed in going round the earth, that is, a month; but whilst she is performing part of her revolution on her axis, she at the same time performs an equal part of her circle round the earth, and thus, by putting herself in a new situation, continues to shew the same side: therefore although with regard to the sun and the rest of the heavenly bodies the moon evidently turns on her axis, yet when viewed from the earth she does not appear to do so. All the other luminaries seem to the moon to rise and set in the space of a fortnight, but she constantly sees our globe in the same part of the heavens.‡ This apparent immobility, were it invariable, would be thought

* This is an error, for if they consider the earth's situation relatively to the firmament, they must see that she performs a revolution in twenty-seven days: they certainly always find her answer to their zenith, or at the same distance from the zenith, but at the same time this zenith is continually answering to some new point in the heavens.

† When the moon is viewed through a telescope, its spots bear no resemblance to the human face; but on contemplating it with the naked eye, it is easy to imagine that form; and it is become so common to talk of the face on the moon, that even an astronomer can hardly divest himself of the idea.

‡ The earth always answers to one side of the moon, but not the same point in the sky.

inconsistent with the nature of a planet; but the moon has a sort
of vibratory motion which sometimes conceals a small part of the
face, and exhibits a part of the other side. Now, I can venture to
say that the inhabitants attribute this motion to us, and imagine
that we vibrate in the heavens, like a pendulum.

All the planets, said the Marchioness, are like us human beings,
who always attribute to others what belongs to ourselves. The
earth says: *it is not I who turn, it is the sun.* The moon says: *it is not
I who vibrate, but the earth:* there is error throughout. I would not
advise you to attempt making any reform, answered I; you had
better consider the remaining proofs of the resemblance which
the earth and moon bear to each other. Figure to yourself those
two globes suspended in the heavens. You know the sun always
enlightens one half of a circular body, whilst the other half remains in the shade. There is then one half of both the earth and
the moon, which is enlightened by the sun, or in other words, in
which it is day, and the other half in which it is night. Observe
likewise that as a ball moves with less force and celerity after it has
struck against a wall from which it flies off to an opposite place,
so the light is weaker when reflected to us from a body that only
receives it. The pale light of the moon is in reality the brilliancy
of the sun, but as we receive it merely by reflection, in coming to
us, it is deprived of its strength. Of course, it shines with much
greater splendour on the moon, and for the same reason the dazzling light received by our globe from the sun, must appear faint,
when reflected back to the moon. That part of the moon which to
us appears luminous during the night, is the side which has day-
light; and the part of the earth which is illuminated by the day,
when turned toward the dark side of the moon, affords equal light
to her. All this depends on the mutual position of the earth and
moon. During the first days of the month, when the moon is not
discernible, she is placed between the sun and us, and proceeding
in the day time with the sun: the luminous side is therefore neces-

sarily turned to the sun, whilst the dark part is towards the earth. We are unable so see the unenlightened side of the moon, but this dark half viewing the part of our globe in which it is day, is assisted by our light, and though invisible to us, has the advantage of seeing the earth as a full moon: it is then to the lunar inhabitants *full earth,* if I may so express myself.* After this, the moon advancing in her monthly round, and no longer between the sun and earth, turns towards us a part of her enlightened half, and that we call the crescent. At the same time that part of the moon which is involved in the obscurity of night, ceases to see all the luminous side of the earth, and finds it continue to decrease.

Enough—said the Marchioness, in her lively manner; I shall easily learn the rest when I like: let me stop a moment, and trace the moon through her monthly circle. I see that in general that planet and the earth have very different degrees of light, and I imagine that when we have the full moon all the luminous side of the moon is turned towards all the part of our globe which is obscure; and that, at that time, the inhabitants cannot discern us at all, but say they have *new-earth.* I should not chuse to be obnoxious to reproach for obliging you to enter into a long explanation of any thing so easily understood, but the eclipses—how are they effected? You could guess it without difficulty, I replied. When we have a new moon, and she, being between us and the sun presents her dark side to our luminous half, the shadow of this obscure part falls on the earth; so that wherever the moon is in a direct line under the sun, she hides that luminary from our sight, and darkens a part of the enlightened side of our globe; this, then, forms an eclipse of the sun to us during the day-time, and an eclipse of the earth to the moon during her night. When the moon is at the full, the earth is between her and the sun, the shaded side of the earth towards the light side of the moon. If the earth's shadow fall

* We have a convincing proof of the light reflected from the earth at this time, in the dusky light perceived on a part of the moon that is not enlightened by the sun. *Astron.,* Art. 1412.

directly on the moon, it darkens the luminous half that we see; 'tis then we have an eclipse of the moon in our night, and the moon, an eclipse of the sun in her day. What prevents an eclipse every time the moon is between the sun and us, or the earth between the sun and moon, is this; it often happens that these three bodies are not placed exactly in a line, in which case the one that would occasion the eclipse throws its shadow on one side of the other, and consequently does not obstruct its light.

I am very much astonished, said the Marchioness, that there is so little mystery in eclipses, and that being produced by such simple means, every body does not discover the cause of them. In truth, answered I, there are many people, who from the emotions they feel at one of the phenomena, appear to have little chance of finding out the occasion of them at present. Throughout the East-Indies, when the sun and moon are eclipsed, the inhabitants believe that a great dragon, with his black claws, is going to seize these luminaries; and all the time the eclipse lasts, you may see whole rivers covered with the heads of these Indians, who have put themselves up to the throat in water, because, according to their notions, this is a very religious act, and will induce the sun or moon to defend itself bravely against the dragon. In America, it was thought that the sun and moon were angry when they were eclipsed, and every kind of absurdity was practised to regain their favour. The Grecians too, who had arrived at such a height of refinement—did they not, for a long time, believe that the moon was eclipsed by the power of sorcery, and that the magicians caused her to descend from the skies, and cast a baneful influence on the herbs? And were not we, likewise, in great alarm but two-and-thirty years ago,[*] at a total eclipse of the sun? Did not an immense number of people shut themselves up in caves and cellars; and were they easily persuaded to leave them by the philosophers who wrote so much to re-assure them?

[*] 1654. There have been others in Europe, in 1724, 1715, and 1716.

Really, replied she, all that is too ridiculous. There ought to be a decree passed to prevent any body from ever talking of eclipses, lest the memory of these follies should be perpetuated. The decree, said I, should extend so far as to obliterate the memory of every subject, for I can think of nothing in the world which is not the monument of some human folly.

Answer me this question, said the Marchioness: Are the inhabitants of the moon as much afraid of eclipses as those of the earth? How ridiculous it is if the Indians of that world put themselves up to the chin in water; if the Americans believe the earth is angry with them; if the Greeks imagine we are enchanted, and suppose we shall injure their herbs; and, in short, if we are inflicting on them all the terror they have caused us? I have no doubt but that is the case, answered I; for why should the good folks in the moon have more sense than we? What right have they to frighten us, unless we can frighten them? I dare say, added I, laughing, that, as a prodigious number of men have been, and still are silly enough to worship the moon; so there are some in the moon that pay their adorations to the earth, and that they are kneeling to one another. If it be so, she replied, we may pretend to have an influence on the moon, and to produce the crisis in the diseases of her sick people; but as a little common sense in the dwellers on that globe would be sufficient to destroy all these honours, I must confess I am afraid they will have the advantage over us.

Don't alarm yourself, said I; 'tis not probable that we are the only fools in the universe. There is something in ignorance that is calculated for general reception, and though I can only guess the character of the people in question, yet I have no more doubt, that, could we form the comparison, we should find ourselves equal to them, than I have that the accounts are true that we receive of their globe.

What accounts do you receive? enquired she. Those, I replied, that are given us by the learned, who travel there every day by the

assistance of telescopes. They tell us that they have discovered in
the moon, earth, seas, lakes, elevated mountains, and profound
abysses.

You astonish me, cried the Marchioness: I can[55] imagine the
possibility of discovering mountains and abysses, from the great
irregularity they cause on the surface of the globe; but how do
they distinguish earth from sea? Because, answered I, the water, *
by suffering part of the light to pass through it, and consequently
reflecting less than the earth, has, at a distance, the appearance
of dark spots; whilst solid parts, by reflecting all the light, look
much more brilliant. The illustrious M. Cassini, who has acquired
a greater knowledge of the celestial bodies than any man in the
world, discovered in the moon something which separates, then
re-unites, and afterwards loses itself in a cavity. We have reason
to believe, from its appearance, that this is a river.[56] In short, all
these different parts are now so well known to us, that they have
been named after our great men. One place is called Copernicus,
another Archimedes, another Galileo.[57] Other parts have fancy
names: there is a promontory of dreams,[58] a sea of nectar, and so
on; in fact, our description of the moon is so particular, that if a
learned man was to take a journey there, he would be in no more
danger of losing himself than I should in Paris.

But, said she, I should like to have a more detailed account of
the interior of the country. The gentlemen of the observatory
are not able to give it you, I replied; you must make enquiry of
Astolfo,[59] who was taken to the moon by St. John. That is one of
the pleasantest follies of Ariosto, I'm sure you will be amused with
it. I confess it would have been better if he had not introduced in it
so respectable a name as that of St. John; poets, however, will take
licenses; and we may venture to excuse this, for the whole poem
is dedicated to a cardinal, and one of our popes has honoured it

* It is proved that there is no water in the moon, but there are volcanoes; they may
even be seen without a telescope, which was the case on the 7th of March, 1794. *Philos.
Trans.*

with a particular eulogium, which in some editions is placed before the work. This is the subject of the piece: Orlando, nephew to
Charlemagne, had lost his senses, because the beautiful Angelica
preferred Medore to him. Astolfo, a valourous knight-errant, was
one day carried by his hippogriffe to the terrestrial paradise, which
was at the top of a very high mountain: there he met with St. John,
who informed him that it was necessary, in order to cure Orlando
of his madness, for them to take a journey together to the moon.
Astolfo, delighted with the opportunity of seeing a new country,
needed no entreaty, and in a moment the apostle and knight took
their course in a chariot of fire. As Astolfo was no philosopher,
he was surprised to find the moon much larger than it appeared
while he was on the earth; his astonishment, however, increased
when he saw in it rivers, lakes, mountains, towns, forests, and
what I should have been equally surprised at, nymphs hunting in
the forests. But the most curious thing of all he saw, was a valley,
in which was to be found every thing that was lost on the earth:
crowns, riches, the rewards of ambition, hopes without number,
all the time that had been devoted to gaming, all the alms that men
had ordered to be distributed after their death, verses dedicated to
monarchs, and the sighs of lovers.

As to lovers' sighs, rejoined the Marchioness; I don't know
what became of them in Ariosto's time, but at present I fancy
there are none that go to the moon. We should find a great many,
said I, were they only those that you have occasioned. In short,
the moon is so careful in collecting all that is lost here, that not a
single thing is wanting of the number; Ariosto has even whispered
that Constantine's donation[60] is there: the popes have assumed
the government of Rome and Italy, by virtue of a donation from
that emperor, but the truth is, we can't tell what is become of it.
There is but one sort of thing that has not escaped to the moon,
and that is—folly: the people on earth have taken care not to part
with that; but to make the moon amends, an incredible quantity

of wit has taken its flight thither, which is there preserved in phials; it is a very subtile fluid, and easily evaporates, unless carefully corked up: on each of these phials is written the name of the owner. I think Ariosto puts them together without any order, but I like better to imagine them placed neatly in long rows. Astolfo was astonished to find full phials belonging to many wise people of his acquaintance. I am sure, continued I, mine has been considerably augmented since I began to indulge myself with you in philosophic and poetic reveries; but I console myself by supposing that, after listening to all my fancies, *your* wits must inevitably become so volatile, that, at least, a little phial full will evaporate, and make its way to the moon.

Our knight-errant found his own among the rest, and by St. John's permission, took possession of it, and snuffed up all the bottleful, like Hungary water; but, according to Ariosto, he did not carry it away with him; for, it soon returned to the moon, in consequence of an extravagance he was guilty of some time after. He did not forget Orlando's phial, which had occasioned his journey; he had a good deal of trouble in carrying it, for the hero's wit was naturally weighty, and not a drop was wanting. At the end, Ariosto, according to his general custom of saying whatever he pleases, addresses in beautiful language, the following apostrophe to his mistress: 'Who, my fair one, will ascend to the heavens, to restore the senses of which your charms has robbed me? Hitherto I have not complained, but I know not what may be the extent of my loss; should I continue the victim of your beauty, I shall in the end become what I have represented Orlando to be. However, I do not believe it is necessary for me to traverse the airy regions for the recovery of my senses; all the faculties of my soul, instead of mounting to such unattainable height, are solacing themselves in the beam of your eyes, and hovering round your lovely mouth. Ah! have compassion on me, and suffer me to take them back with my lips.' Is not the thought pretty? For my part, in adopting Ariosto's way of thinking, I should dissuade people from ever letting their wits escape, unless it were from the influence of love; for you see how near they then continue, and how easily they may be regained; but when they are lost in any other way, as we, for instance, are losing ours in philosophising, they fly directly to the moon, and are not caught again at pleasure. Never mind, said the Marchioness; ours will have an honourable station among the philosophic vials; whereas, had we lost them in the poet's way, they might, perhaps, hover around some unworthy object. But, continued she, to deprive me completely of mine, tell me, seriously, whether you believe there are men living in the moon, for you have not yet given me a decided opinion. Do I believe it? replied I; Oh no, I don't believe there are men in the moon. We see how much all nature is changed even when we have travelled from here to China; different faces, different figures, different manners; and almost a different sort of understandings: from here to the moon the alteration must be considerably greater. When adventurers explain unknown countries, the inhabitants they find are scarcely human; they are animals in the shape of men, even in that respect sometimes imperfect; but almost devoid of human reason;[61] could any of these travellers reach the moon, they surely would not find it inhabited by men.

Then what sort of creatures are they? asked the Marchioness, impatiently. Upon my word, Madam, said I, I can't tell. Were it possible for us to be endowed with reason, and at the same time not of the human species; were we, I say, such beings, and inhabitants of the moon, should we ever imagine that this world contained so fantastical a creature as man? Could we form in our minds the image of a being composed of such extravagant passions, and such wise reflections: an existence so short, and plans so extensive; so much knowledge of trifles, and so much ignorance of the most important things; such ardent love of liberty, yet such proneness to slavery; so strong a desire for happiness, with

so little power of being happy? The people in the moon muſt be
very clever to imagine such a motley charaɛter. We are incessant-
ly contemplating our own nature, yet we are ſtill unacquainted
with it. Some have found it so difficult to comprehend, that they
have said the gods had taken too much neɛtar when they created
men; and when they had recovered their calm reason, they could
not help laughing at their own work. Well, we are not in danger
of being laughed at by the inhabitants of the moon, answered the
Marchioness, as they would find it so impossible to imagine our
charaɛters: but I should be very glad if we could find out theirs;
for, really, one feels a painful degree of curiosity in knowing that
there are beings in the moon we see yonder, and not having the
means of discovering what they are. How is it, I replied, that you
have no anxiety to be acquainted with all the southern parts of the
world, which is yet unknown to us?[62] we and the inhabitants of
that part of the globe are voyaging in the same vessel, of which
they occupy the head, and we the ſtern. You see that the head and
the ſtern have no communication with each other; that the people
at one end know nothing of the nature or occupations of those at
the other, and yet you want to be acquainted with all that is going forward in the moon, that separate vessel which is sailing in a
diſtant part of the heavens.

Oh! replied she, I consider myself already acquainted with the
inhabitants of the southern world, for they certainly muſt be very
much like us; and in short, we may know them better whenever we
chuse to give ourselves the trouble of going to see them; we cannot
miss them, for they will remain in the same place; but these folks in
the moon—I am in despair about them. Were I, I replied, gravely
to answer you, *we know not what may happen,* you would laugh at
me, and I should undoubtedly deserve it; nevertheless, I think I
could defend myself, in some measure, from your ridicule. A
thought has come into my head, which is whimsical enough, and
yet there is a wonderful deal of probability in it; I don't know how

it has acquired the power of imposing that on my understanding, being in itself so extravagant. I dare say I shall likewise bring you to confess, contrary to reason, that there may some day be a communication opened between the earth and the moon. Recollect the situation of America before it was discovered by Christopher Columbus. The minds of its inhabitants were involved in the most profound ignorance, far from having any knowledge of the sciences, they were not even acquainted with the most simple and necessary arts; they went without clothes; they had no weapon but the bow; they had no notion that men might he carried by animals; they supposed the ocean an immense space, impassable by man, and bounded only by the sky, to which it was joined. It is true, that after they had been several years in contriving to scoop out the trunk of a great tree, they ventured to commit themselves to the water in this rude sort of vessel, and went from one country to another, borne along by the winds and waves: but as their bark was very liable to be overset, they were frequently under the neccessity of swimming to overtake it, so that, properly speaking, they were oftener in the water than in their ship. You must suppose they would not have yielded a very implicit credence to a person who had told them that a navigation was carried on, incomparably superior to theirs; that by its means, every part of the liquid expanse could be resorted to; that the vessels might be detained at one spot, whilst the billows were foaming around; that even the speed with which they moved might be regulated; in short, that the ocean, whatever its extent might be, was no obstacle to the commerce of different people. In a course of time, however, notwithstanding their incredulity, a spectacle new and astonishing presents itself to the eyes of these savages. Enormous bodies, extending their white wings to the blast, come sailing on the ocean with fearful rapidity, and discharging fire on every side: these tremendous machines cast on their shore men covered with iron; guiding with facility the monsters that carry them, and darting thunderbolts from their hands, to destroy all who attempt to resist them.——'Whence come

these awful beings? Who hath given them power to ride on the
waters and to wield the thunder of heaven? Are they children of
the sun? assuredly they are not men!' I cannot tell, Madam, whether you feel as ſtrongly as I do, the surprise of the Americans; surely no event could ever have excited an aſtonishment equal to theirs.
After thinking of that, I will not assert that no communication can
be eſtablished between our world and the moon. Did the Americans ever conceive the idea that there would be any between their
country and Europe, of which they had never heard? There is, I
acknowledge, an immense space of air to travel through before we
could reach the moon; but did those great seas appear to the Americans more capable of being crossed? Really, exclaimed the Marchioness, looking earneſtly at me, you are quite mad! Who denies
it? answered I. It is impossible you *should* deny it, said she. The
Americans were so ignorant, that they could not imagine the prac-
ticability of crossing such an extent of water; but we have science
enough to know that the air is passable, although we have no ma-
chine which can transport us through it. We do more than conjecture the possibility of rising in the air, I replied, we have actually
begun to fly. Several persons have discovered a method of fixing
on wings which supported them in the air, of moving these wings,
and by their assiſtance, flying over rivers; these new-fashioned
birds did not, to be sure, soar like the eagles, and their flight has
sometimes coſt them an arm or a leg; but, however, these attempts
answer to the firſt pieces of wood that were launched into the water, and which served for the commencement of navigation: there
was a vaſt difference between these mere planks and great ships,
capable of going round the world; nevertheless, by gradual improvements we have learned to conſtruct such vessels. The art of
flying is but in its infancy; in due time it will be brought to
perfection,*and some day or other we shall get to the moon. Can
we pretend to know everything? to have made every possible dis-

* Montgolfier's balloons, invented in 1783, have gone a great way in fulfilling of this
prediction, but it is evidently impossible for it to be accomplished; these globes can only
carry us to a certain height, beyond that we could not breathe.

covery? Pray let us give poſterity leave to make some improvements as well as ourselves. I won't give them leave, answered she, to break their necks by attempting to fly. Well, I replied, though flying be not perfeċted here, the inhabitants of the moon may, perhaps, excel us; and it will be the same thing, whether we go to them, or they come to us. We shall then be like the Americans, wbo knew so little of navigation, while it was so thoroughly underſtood at the other side of the globe. Pugh! cried the Marchioness; if the people of the moon were so expert, they would have been here before this time. The Europeans, answered I, did not find their way to America till six thousand years had elapsed; they were all that time learning the art of navigation so completely as to pass over the ocean. Probably the people in the moon are able to take little excursions into the air, very likely they are now praċtising; after they have acquired more experience they will pay us a visit, and heaven knows what surprise it will occasion us! You are insupportable, exclaimed she, to combat me with such chimerical arguments. Take care, said I; If you provoke me I shall corroborate them. Remember the earth has been made known to us by little and little. The ancients positively asserted that the torrid and frozen zones were uninhabitable, from the excessive heat of the one, and the cold of the other; and in the time of the Romans the general chart of the world was made little larger than that of their own empire, this at once shewed the grand idea they had of themselves, and their extreme ignorance of the earth. Men were, however, discovered in these extremely hot and intensely cold climates, which discovery has greatly augmented the number of inhabitants on our globe. At one time it was believed that the ocean covered every part of the earth except what was then known. Antipodes had never been heard of, and who could imagine that men would be able to walk with their heads downwards ? Yet after all, the antipodes were found out. Now the map muſt be altered; a, new half added to the earth! You underſtand, Madam, what I am aiming at; these antipodes, so unexpeċtedly discovered, should teach us to think

The Montgolfier brothers' first flight in 1783.

modestly of our attainments: we may yet know much more of our own world, and then become acquainted with the moon; till that time we must not expect, because our knowledge is progressive: when we understand our own habitation, we may be permitted to study that of our neighbours. In truth, said she, viewing me attentively, you enter into the subject so deeply that one cannot but imagine you in earnest. Indeed I am not, answered, I; I only wished to shew you the possibility of maintaining an extravagant opinion, so as to embarrass, though not convince, a person of sense. Truth alone makes her way to the understanding; she can even convince without exhibiting every proof: she is so adapted to our capacities, that when first discovered, we seem only to have met with an old acquaintance.

Ah! this restores my tranquillity, said she. Your sophistry disturbed my imagination. Let us retire; I am now composed, and inclined to go to rest.

THIRD EVENING

Dangers Faced by the
Inhabitants of the Moon

THE MARCHIONESS WISHED to pursue our astronomical researches during the day; but I told her that as the moon and stars were the subject of our whimsical conversations, they ought to be our only confidants: we therefore waited till evening, and then took our ramble in the park, which thus became sacred to learning

I have a vast deal of news to tell you, said I: I yesterday told you, that the moon, according to all appearances, was inhabited; but I have recollected a circumstance which would expose its inhabitants to so much danger, that I don't know whether I shall not retract my former opinion. Indeed, I will not suffer you to retract it, answered she. Yesterday you prepared me to receive a visit from the inhabitants of the moon in a few days; now you are going to refuse them a place in the creation. You shall not trifle with me in this way. You told me the moon was inhabited; I surmounted the difficulty of believing it, and now I will continue to believe it. Softly! said I; we should give but half an assent to an opinion of this nature, and reserve the other half in case we should find the opposite idea better supported. I am not contented with words, she replied; give me facts; remember your comparison of the moon with St. Dennis. But, answered I, the moon is not so similar to the earth, as St. Dennis is to Paris. The sun draws out of the earth watery exhalations, which rise to a certain height in the air, collect together, and form themselves into clouds. These clouds hover about the earth in irregular shapes, sometimes shadowing one part, and sometimes another. In viewing earth from a distance, the appearance of its surface would continually vary, because a large space of country darkened by a cloud would appear less luminous than the other parts, and as the cloud dispersed, would resume its brightness: from this cause the spots on the earth would be seen to change their places, assume different forms, and sometimes be entirely dissipated. If, then, the moon had clouds in its atmosphere, we should observe this variety of spots; but we find them always confined to the same place, which proves that the sun raises no vapours from the moon. It is then a body incomparably more solid than the earth, and its subtile particles easily dissipated as soon as they are put in motion by the heat. The moon, therefore, must be a mass of rock and marble, from which no evaporation proceeds; for exhalations so naturally arise where there is water, that we cannot admit the existence of water where they are not to be found. What sort of beings do you think could inhabit these barren rocks; this country without water? Ah! cried she, you forget that you have assured me the seas in the moon were distinguishable. It was a mere conjecture,* I replied; I am sorry to have led you astray. These dark places which have been taken for seas are probably only deep cavities: at so great a distance it is excusable if we don't always guess aright. But, said she, will your objections oblige us to conclude that the moon has no inhabitants? By no means, answered I, we will neither decide one way or another. I must own my weakness, she replied; I cannot bear to remain in suspense; I must believe something; enable me to determine; let us ascertain the existence of these people, or let us annihilate them at once, and think no more about them. But preserve them if possible; I have formed an attachment for them, of which I shall not easily divest myself. I will not leave the moon without inhabitants then, said I; for your pleasure it shall be repeopled.

---

* This is not, now, even conjectured, for with a telescope we may see irregularities at the bottom of what were supposed to be seas.

As the spots in the moon never vary,* we certainly cannot believe that there are any surrounding clouds which successively
obscure the surrounding parts; this however is not a proof that
there are no exhalations; our clouds are formed of vapours, which
at their first rising out of the earth, were in separate particles, too
small to be visible to us; in ascending they meet with a degree of
cold that condenses and unites them into conspicuous forms; after
which they float in the air till they dissolve in rain. But these exhalations frequently remain dispersed and imperceptible, and fall
back on the ground in gentle dews. I suppose then that vapours
of this kind are exhaled from the moon, for it is incredible that the
moon should be a large mass, composed of parts all equally solid,
all in a state of equal tranquillity, all incapable of being influenced
by the action of the sun. We know of no body which has these
properties, not even marble. The most dense bodies are subject
to change, either from some secret and interior motion, or from
the action of external matter. As the exhalations from the moon,
do not form themselves into clouds, and return in showers, they
can only become dew: for that purpose it is not necessary that the
atmosphere, which apparently adheres to the moon as ours does
to the earth, should be exactly similar to our air, nor the vapours
exactly like ours; and that I think is probably the case:† the matter must have a different disposition in the moon, from that in the
earth; consequently the effects be different; however, all that is
of no importance; since we find that there is motion in the parts of
the moon, either internal, or produced by foreign causes, we may
again people it, as we have the means of affording them subsist-
ence: of producing fruit, corn, water, and every thing that is needful. I mean fruit, corn, &c. such as the moon can produce, the nature of which I am unacquainted with; and all these in proportion
to the wants of its inhabitants, of which I am likewise ignorant.

* M. Herschel has observed variations in them; which he, certainly, attributes to the
industry of the inhabitants.

† The atmosphere of the moon, if there be any, is quite invisible to us.

Sir William Herschel's 40 ft telescope.

That is to say, answered the Marchioness, you are sure every thing is right, without knowing how it is; here is a little knowledge placed againſt a great deal of ignorance, but we muſt be content with it: I am very happy to have inhabitants reſtored to the moon; I am glad also that you give them a surrounding atmosphere, for it seems to me that a planet would be too naked without one.

These two different airs, said I, one belonging to the earth, the other to the moon, tend to prevent a communication between the two planets. If it merely depended on the power of flying, who knows, as I yesterday said, but we may, at some future time, be sufficiently expert? All things considered, I think we must not expect this communication; the amazing distance at which they are placed, would be a considerable difficulty; and were this obstacle removed, were the two planets nearer together, it would be impossible to pass from one atmosphere to the other. Water is the atmosphere of the fishes; they never pass into that of the birds, nor the birds into theirs: they are not prevented by the distance, but the existence, of both depends on their proper element. Our air, we find, is mixed with more dense and gross vapours than that of the moon; therefore an inhabitant of that world would be drowned if he entered our atmosphere, and fall lifeless on the earth.

Oh! how glad I should be, exclaimed the Marchioness, for a shipwreck to cast a good number of them on the earth, we might then examine them at our leisure. But, I replied, if they were clever enough to navigate the surface of our atmosphere, and from a curiosity to examine us, should be tempted to draw us up like fishes, would that please you? Why not, answered she laughing. I would voluntarily put myself in their net, just for the pleasure of seeing the fishers.

Remember, said I, you would be very ill by the time you reached the top of our air; we are not capable of breathing it above a certain height:* it is said that at the summit of some mountains we can scarcely do it. I wonder that people who are silly enough to believe that corporeal genii inhabit the purest regions of the air should not tell us, as the reason for our receiving such short and unfrequent visits from these genii, that few of them understand diving, and even those who excel in it cannot remain long in our gross air.

We see then there are many things to prevent us from leaving our own world, and going to the moon. To console ourselves let us guess all we can about it. In the first place, I conjecture that

---

* Respiration is difficult at the height of a league. Half a league higher it must be impossible.

the inhabitants muſt see the heavens, the sun, and the ſtars, of a very different colour from what they appear to us. We view those objeƈts through a sort of glass which alters their appearance; this glass is our atmosphere, pervaded with exhalations. Some moderns assert that it is blue, as well as the sea, but we can only diſtinguish the colour in the parts of those elements that are moſt remote from the eye. The firmament, say they, in which are the fixed ſtars, has no light in itself, and consequently ought to appear black,* but as we see it through our blue air, it seems to us to be blue. If that is true, the rays of the sun and ſtars cannot pass through the air without receiving a slight tinge from its colour, and losing a degree of that which is natural to them. But supposing the air is not coloured, it is certain that through a thick fog the light of a flambeau, seen at some diſtance, appears of a deep red, which is not its real colour; if therefore our air be considered only a miſt, it muſt necessarily alter the colour of the sky, sun and ſtars. The celeſtial fluid alone could give us light and colours in their original ſtate. Therefore as the atmosphere of the moon differs from ours, it is either of a different colour, or else it is another sort of miſt, which varies the appearance of the celeſtial bodies. In a word, the glass through which the people in the moon view these objeƈts is of a different nature to ours.

On that account, replied the Marchioness, I prefer our world to the moon; I think it impossible for the assortment of colours presented to our sight by the heavenly bodies to be so beautiful as that they form when viewed through the medium of our air. Let us suppose a red sky and green ſtars; the effeƈt is not so agreeable as golden ſtars and a blue sky. One would think, said I, you were chusing clothes or furniture; but believe me, nature has a good taſte; let us truſt to her for providing a set of colours for the moon; there is no fear but it will be a pleasing one. She has un-

* De Saussure tells us it appears black when viewed at a league's diſtance from the earth.

doubtedly varied the appearance of the universe at each different point of view, and in all these varieties there is great beauty.

I acknowledge her talents, answered she; at each point of view she has placed a different sort of glass, by which mean she has given the appearance of variety to objects which remain always the same. With a blue atmosphere, we have a blue sky, and perhaps with a red atmosphere, the inhabitants of the moon have a red sky; yet this sky is absolutely the same. In like manner she Seems to have placed various sorts of glasses before the eyes of our imagination, through which the same object presents to each of us a different appearance. To Alexander, the earth appeared a proper place to convert into an empire, for his sway; Celadon viewed it only as a fit residence for Astrea;[63] a philosopher considers it a large planet travelling through the heavens, and inhabited by a number of madmen. I think the spectacle of nature cannot be more varied than the prospects of different imaginations.

The varied appearance of objects viewed by the imagination, I replied, is the most surprising, for they are exactly the same things though apparently so dissimilar: whereas there may be other natural objects visible to the moon, and some that are visible to us may not be seen there; perhaps, for instance, there is neither dawn nor twilight. The air that surrounds us, rises to some height, receives the rays of light that would not reach the earth, and by its density, detains, and conveys to us a part of this light which was apparently not destined for us: thus you see the dawn and twilight are particular favours conferred on us by nature; they are degrees of light to which we are not regularly entitled, and which are bestowed on us in addition to our share. But the atmosphere of the moon, being purer than ours, is probably not so well calculated to reflect the rays which it receives before the sun is risen, or after it is set. The poor inhabitants have not then this light, which by its gradual increase prepares us so agreeably for the brilliancy of the sun; and in the evening reconciles us to its loss, by a progressive diminution. The moon, after the profound gloom of night, receives the ardent blaze of the sun, as if by the instantaneous drawing up of a curtain: on the contrary, whilst still enjoying the dazzling light of day, it is again plunged into extreme darkness: day and night are not connected by an agreeable medium, partaking of both. The people in the moon never see the rainbow; for as the dawn is produced by the thickness of our air and vapours, so the rainbow is formed in the clouds which are dispersed in rain: thus we are indebted for the most beautiful appearances in nature, to things, in themselves, far from agreeable. Since the moon has neither dense vapours nor rainy clouds, farewell to Aurora, and the Rainbow! Alas! to what can they liken the beauties of that country; what a source of comparison are they deprived of!

I should not much regret those comparisons, answered the Marchioness, and I think the inhabitants of the moon have ample amends made them for the loss of rainbows and twilight by being exempted from thunder and lightning; for these likewise are formed in the clouds. They have constant serenity of weather; never losing sight of the sun. They have no gloomy nights in which the stars are concealed. They are unacquainted with those storms and tempests; those elemental wars which seem to indicate the wrath of heaven. Are they then to be pitied ? You speak of the moon as an enchanting spot, said I; yet I don't know whether it is very delightful to be exposed throughout a day that is as long as our fortnight* to a blazing sun, without a cloud to temper the intensity of its heat. It is perhaps owing to this that nature has formed cavities in the moon, large enough to be seen by our telescopes; they are not valleys situated between mountains, but hollow places in the midst of large planes. How do we know whether the inhabitants, oppressed by the perpetual radiance of the sun, may not take refuge in these caverns? Perhaps they even build towns, and constantly reside in these parts. We see that here

* During this time the sun rises and sets as it does in our day.

our subterraneous Rome is larger than the Rome which is built on the surface: we have only to remove the latter, and the other would be a city such as we should find in the moon. A large number of the people dwell in each cavern, and from one cavern to another is a subterraneous passage for the communication of the inhabitants. You laugh at this idea; I have no objection; but seriously, I think you are more likely to be miſtaken than I. You believe the people in the moon muſt dwell on the surface, because we are on the surface of our globe; you should form quite a different opinion and think that because we reside on the surface they dwell in the interior parts; every thing muſt be very differently conducted here and in the moon.

It does not signify, replied the Marchioness; I can't bear the idea of those people living in perpetual darkness. You would find it ſtill more difficult to admit the opinion, said I, if you knew that a great philosopher of ancient times had informed us that the moon was the dwelling of souls who had on earth rendered themselves worthy of very exalted happiness. He supposes that their felicity consiſts in liſtening to the music of the spheres, but that when the moon comes under the shadow of the earth, they are no longer able to hear the celeſtial harmony, at which time they utter the moſt piercing cries, and the moon haſtens on as faſt as possible to relieve them from this agonizing situation. We may expect then, answered she, to have the virtuous spirits sent here from the moon, for I suppose they likewise honour our world by making it an abode of the blessed; so in these two planets it is thought a sufficient reward to superior goodness for the soul to be transported from one world to the other. Really, I replied, it would not be a trifling enjoyment to take a survey of different worlds; I often receive a great deal of pleasure from such a journey, although but in imagination; what muſt it be then to perform it in reality? It would be much more delightful than going from here to Japan; in other words, than crawling from one end of the earth to the other

with great labour, merely for the sake of seeing men. Well, said
she, let us make this tour to the planets as we can; what should
prevent us; we will place ourselves at all those different points of
view, and at each of them survey the universe. Have we any thing
else to see in the moon? You are not yet thoroughly acquainted
with that world I replied. You recollect that the two motions of
the moon, by one of which she turns on her axis, and by the other
round us, being equal, the latter always prevents the former from
withdrawing any part from our sight, and consequently we always view the same side. That half therefore is the only part that
can see our world, and as the moon, with regard to us, must be
considered not to turn on her centre, the half to which we are visible, sees us always fixed in the same part of the sky.* When it is
night, and the nights there are as long as our fortnight, she sees at
first only a small part of the earth enlightened; then a larger portion, and at length the light seems hourly to spread over the earth,
till it becomes entirely luminous. On the contrary these changes
in the moon are visible to us only from one night to another, because we are a long time without seeing her, I should like to hear
the mistakes which the philosophers of that world fall into from
the apparent immobility of our earth, whilst all the other heavenly
bodies rise and set in the space of a fortnight. Probably they consider the earth immoveable in consequence of her enormous size,
being sixty times larger than the moon; and when the poets are
disposed to flatter indolent princes, I have no doubt but they compare them to this orb in her state of majestic repose. It does not
however appear an intire immobility. From the moon they must
see the earth turn on her axis. Our Europe, Asia, America, present
themselves one after another, in different shapes, nearly as they
are represented on our maps. Only imagine what a novel sight this
must be to travellers coming from the other side of the moon to
that which is always facing us! How incredulously they must have

*   That is to say only at the same distance from the zenith and the horizon.

heard the accounts of the firſt that spoke of it, who lived at the opposite side. It is come into my head, said the Marchioness, that from that half of the moon to the other they make pilgrimages to come and examine us, and that particular honours and privileges are deſtined for those who have seen the great planet. At leaſt, answered I, they who conſtantly see us have the privilege of being better illumined during their nights; the inhabitants of the other side muſt be much less agreeably situated in that respeĉt.

Now, madam, let us pursue our journey to the different planets; we have been long enough at the moon. Next, in the road from the moon to the sun, we find Venus. In talking of Venus, I shall resume my argument concerning St. Dennis. Venus, as well as the moon turns on her axis and goes round the sun: with telescopes it is seen that this planet, like the moon, is sometimes a crescent, sometimes on the decrease, sometimes full, according to her different situations relatively to the earth. The moon, according to all appearances, is inhabited; why then should Venus be deſtitute of inhabitants? But, interrupted the Marchioness, with your *why nots* you will put inhabitants in all the planets. Certainly, I replied, this *why not* has the power of peopling them all. We find that they are of the same nature, all opaque bodies, illumined only by the sun, and the refleĉtions of his rays on each other; and having all the same motions. So far then they are alike, and yet we are to suppose that these great planets were formed to remain uninhabited, and that such being the natural condition of them all, an exception should be made in favour of the earth——let who will believe it; I cannot. A few minutes, answered she, have wonderfully confirmed your opinion. Juſt now the moon was on the point of being quite deserted, and you cared very little about the matter, and now, if one were to presume to deny that all the planets are as full of inhabitants as the earth, I see you would be quite in a passion. It is true, said I, that in the positive fit I had juſt now, if you had contradiĉted me on the subjeĉt of these said inhabitants, I should not only have maintained their exiſtence, but in all probability have described their formation. There are certain moments when we feel assured of a thing, and I never felt so fully persuaded of my opinion as I was then; however, though my ardour is now a little abated, I still think it would be very strange for the earth to be so well inhabited, and the other planets perfectly solitary; and numerous as we know the inhabitants of the earth to be, we do not see them all; our world contains as many species of animals that are invisible to us, as of those that we discern. From the elephant to the hand-worm we can examine them; there our sight is bounded; but after the hand-worm is an infinitude of little animals not discernible by the naked eye, and to which, in point of size, he is an elephant. With magnifying glasses, we may see a drop of water, vinegar, or any other liquor, filled with little fishes or serpents, which we should never have thought of finding there;[64] and some philosophers suppose the taste of these liquors is produced by punctures which the little animals make in the tongue. Mix these liquors with certain things, expose them to the sun, or leave it to corrupt, and you will find new sorts of animals.

Many masses, apparently solid, contain scarcely any thing but a heap of these small animals, which in so confined a situation find room enough for their little movements. The leaf of a tree is a world, inhabited by worms imperceptibly small, to which it appears an amazing extent, having mountains and caverns, and so large that from one side of the leaf to the other the little worms have no more communication with each other than we have with the antipodes. From such considerations I cannot doubt of a great planet being inhabited. There have been found even in very hard stones an endless number of worms lodged in every interstice, feeding on parts of the stone. Consider the countless numbers of these little beings, and how many years they could subsist on a quantity of food as big as a grain of sand; and then though the moon should be but a mass of rock, we may let it be eaten by its inhabitants rather than not assign any to it. In short every thing is animated; every thing is full of life. Associate in your calcu-

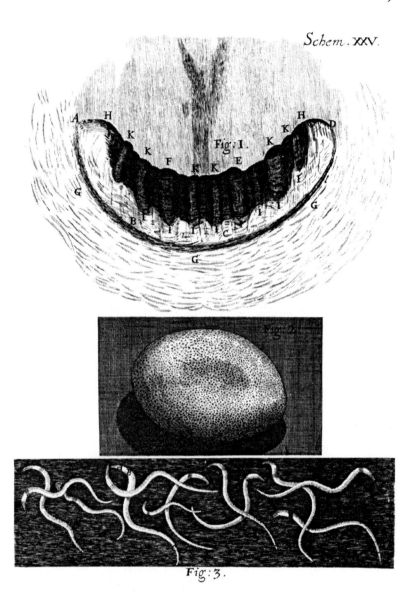

Microscopic view of (top) a snail's teeth, (centre) a silkworm's egg and (bottom) 'eels' in vinegar, from Hooke's *Micrographia*, 1665.

lation all the species that have been lately discovered, and that those we may suppose are yet undiscovered, with all that we are in the habit of seeing, and you will surely confess that the earth is amply stocked with living creatures; that nature must delight in bestowing life since she has created such infinite variety of beings so small as to elude our sight. Can you believe that after the earth has been thus made to abound with life, the rest of the planets have not a living creature in them?

My reason is convinced, answered the Marchioness; but my imagination is overwhelmed with such an infinite variety and number of inhabitants existing in each of the planets; for as there is no dull uniformity in nature, the difference of species must be in proportion to the number of beings—how can imagination grasp such a vast idea? Imagination, I replied, is not required to represent all this to us; we can penetrate no farther than we are assisted by our sight; we can only perceive, from a general glance, that nature has established an inconceivable diversity in her works. The human face is formed everywhere on the same plan, but still how great is the difference between the visages of Europeans and of Africans or Tartars: not only in separate nations do we find a distinguishing character of countenance, even among the same people every family seems formed from a distinct model. How astonishing is the power of nature in giving such variety to so simple an object! In the universe we are but as a little family whose faces resemble each other; the next planet contains another family who have a different style of countenance. Probably the variations are greater in proportion to the distance, and could we compare the inhabitants of the earth and moon, we should easily see that they were nearer neighbours than those of the earth and of Saturn. Here, for instance, our thoughts are made vocal; the people in another planet only express themselves by gestures; farther off, they may dispense with any sort of conversation. Here our reason is matured by experience; elsewhere experience may add little to the understanding; at a greater distance, children may know as much as old men. In this world we give ourselves more uneasiness about the future than the past; on another globe, the past afflicts more than the future; on a third, the people are neither distressed by one nor the other, and they perhaps are not the most unhappy. It is said that we are possibly in want of a sixth sense belonging to our nature, by means of which our knowledge would be greatly augmented. This sense is most likely in some other world, where one of our five is wanting. There may even be a great number of natural senses, but, in the distribution of them among the planets, only five have fallen to our share; and with these five we remain satisfied because we don't know of any more. Our sciences have certain limits which no human understanding has exceeded: at a particular point we stop, the rest is reserved for other worlds, when they are ignorant of many things that we know. This planet is blest with the delightful emotions of love, but at the same time desolated by the fury of war. Another enjoys perpetual tranquillity, but with this uninterrupted peace, love is unknown, and calmness degenerates into ennui. In short whatever nature has done on a small scale, for the distribution of happiness and talents among us, she has undoubtedly performed on a more extensive plan for the benefit of the universe; at once diversifying and equalizing all.

Are you satisfied, madam, said I? Have I given your imagination room to exert itself? Do you not already see the people of different planets? No, answered she, with a sigh: all you have been saying is so vague and unsatisfactory; there is nothing in it for the mind to fix on. I want something more determined; more marked. Well then, I replied, I will not conceal any particulars that I am acquainted with: I can give you some information that you will acknowledge to be undoubted, when I tell you my authorities. Prepare to listen patiently if you please, for it is a long story.

In one of the planets, I shall not at present tell you which, there is a people that are very active, laborious and skilful. Like some of

our Arabs, they live by pillage, and that is their only fault. They
live together in the moſt harmonious manner, labouring inces-
santly and in concert, for the common good: above all their chas-
tity is unexampled; it is true they have no great merit in it; they
are all ſterile; there is no difference of sex among them. But, interrupted the Marchioness, were you not aware that the author of
this marvelous ſtory wanted to make a fool of you? How could
such a nation be perpetuated? No, I replied, very coolly, they did
not intend to make a fool of me; all that I have told you is faɛ̃t, yet
the nation is perpetuated. They have a queen, whose royalty consiſts, not in direɛ̃ting the business of the ſtate, nor in leading her
subjeɛ̃ts to the field of battle, but in her surprizing fecundity, she
has millions of children; in short the produɛ̃tion of them occupies
the whole of her time. She has a large palace, divided into a vaſt
number of chambers, in each of which a cradle is prepared for a
little prince, and she is confined successively in all these chambers,
always surrounded by her courtiers who congratulate her on the
noble privilege she enjoys exclusively of her subjeɛ̃ts.

I see, madam, that you wish to enquire who are her lovers, to
give them a more respeɛ̃table appellation, her husbands. Some of
the eaſtern queens have seraglios of men; she apparently does the
same, but she keeps it a greater secret than they; this may arise
from modeſty, but it is aɛ̃ting with little dignity. Among these
Arabs who are always in aɛ̃tion, are found a few ſtrangers, in person very much resembling the natives of the country, though extremely different in disposition, for they are remarkably indolent;
they never ſtir out nor engage in any business; and were not these
persons kept for the pleasure of the queen, they would hardly be
suffered to remain amongſt so induſtrious a people. If, in reality,
notwithſtanding the smallness of their number, they are the fa-
thers of many thousands of children, they deserve to be excused
from any other employment; and it is a ſtriking proof that this is
their only funɛ̃tion, that as soon as the queen has brought forth

her ten thousand children, the Arabs kill, without mercy, the un-happy foreigners, then become useless to the ſtate.

Have you done? enquired the Marchioness. Thank heavens! Let us now resume a little common sense, if we can. Where have you picked up this romance? What poet is the inventor of it? I again tell you, answered I, that it is no romance. All this takes place on our globe, even under our eyes.—If I muſt explain the myſtery. These Arabs are no other than bees.

After this I gave her the natural hiſtory of bees, of which she had before scarcely ever heard more than the name. In conclud-ing, you see, said I, that in attributing to other planets what is daily passing here, we should be accused of telling the moſt extravagant falsehoods. The hiſtory of inseĉts, in particular, is a colleĉtion of wonders. I have no doubt of it, she replied: the silk-worm alone, with which I am better acquainted than the bees, would afford abundant materials for your descriptions. A people undergoing such wonderful changes as to be totally unlike what they formerly were; at one part of their lives crawling, at another, flying: in short a thousand incredible things might be told of the charaĉter and manners of this nation.

My imagination, continued the Marchioness, is begining to work on the subjeĉt you have given me——the inhabitants of all the planets: I am conjeĉturing their figures; I can discern some of them very diſtinĉtly, but I don't know how to describe them to you. As to their figures, said I, I advise you to leave the forma-tion of them to your dreams, we shall hear tomorrow what they have suggeſted, and whether they have been able to represent the inhabitants of any of the planets.

FOURTH EVENING

Particulars Concerning the Planets
Venus, Mercury, Mars, Jupiter, and
Saturn

THE DREAMS OF the Marchioness did not assist her; they represented nothing that did not bear a resemblance to what we see here. I had the same complaints to make as certain people, whose paintings are always fanciful and grotesque, do at the sight of our pictures. *Phsaw*, say they, *these are all men; here are no objects of imagination*. We therefore resolved to content ourselves with the conjectures we should be able to make concerning the inhabitants of the planets as we continued our journey: we had last night got as far as Venus. We are assured, said I, that Venus turns on her axis, but it is not ascertained in how long a time, consequently, we cannot tell the length of her days.[65] Her year lasts but about eight months, as she is not longer than that in performing her revolution round the sun. She is of the same size as the earth,[66] therefore the earth and Venus appear equally large to each other. I am glad of that, said the Marchioness; then I hope the earth is to Venus the shepherd's star, and the parent of love, as Venus is to us. These appellations can be proper only for a pretty little, brilliant, gay looking planet. True, answered I; but do you know what makes Venus look so beautiful at a distance? it is the effect of her being very frightful when near. With good telescopes it has been seen that she is covered with mountains, much higher than ours, sharp pointed, and apparently very dry.* This kind of surface is the best calculated to reflect the light with great brilliancy. Our earth, whose surface is very smooth, compared with

---

* M. Herschel's observations contradict this idea. Venus has a very dense atmosphere, which prevents us from distinguishing anything on her surface; the brilliant appearance of this planet arises from her proximity to the earth.

that of Venus, and partly covered with water, probably looks less beautiful at a diſtance. So much the worse, said the Marchioness, I should like her to preside over the loves of the inhabitants of Venus; they muſt certainly underſtand what love is. Oh! undoubtedly, I replied; the people in that planet are all Celadons and Sylvanders,[67] and their everyday conversations are finer than the moſt admired in Clelia.[68] Their climate is very favorable to the tender passion. Venus is nearer to the sun than we; and receives more light and heat: she is about two thirds the diſtance of the earth from the sun.

I can see, interrupted the Marchioness, what sort of people the inhabitants are. They are much like the Moors of Grenada: a little, dark, sun-burnt people, scorched by the sun; full of wit and animation, always in love, always making love, liſtening to music, having galas, dances, and tournaments. Give me leave to tell you, Madam, answered I, that you know but little of the inhabitants of Venus. Our Moors of Grenada when compared with them would appear as cold and ſtupid as Greenlanders.

But what muſt the inhabitants of Mercury be? We are above twice the diſtance from the sun that they are. They muſt be almoſt mad with vivacity. Like moſt of the negroes, they are without memory, never reflecting; acting by ſtarts and at random: in short, Mercury is the bedlam of the universe. The sun appears there nine times larger than it does to us; the light they receive is so brilliant, that our fineſt days would be but twilight in comparison; perhaps they would find them so dark, as not to be able to diſtinguish one thing from another. The heat[69] to which they are accuſtomed is so intense, that they would be almoſt frozen in our Africa. In all probability, our iron, silver, and gold, would be melted in their world, and only be seen in a liquid ſtate, as we in general have water, which in some degrees of cold becomes a solid body. The inhabitants of Mercury would not imagine that in another world those liquors, which, perhaps, form their rivers, are the hardeſt of all bodies. Their year laſts but three months. The length of their

day is not known to us,[70] because Mercury is so small, and so near
the sun, that it exceeds the art of all our astronomers to observe
him with sufficient accuracy to determine what sort of motion he
has on his centre; the inhabitants, I think, must wish it to be performed in a short time, for scorched as they are with the fierceness
of the sun, the coolness of night is undoubtedly very desirable to
them. The part which by rotation is deprived of the sun's light, is
illumined by Venus and the earth, which must appear very large.
As to the other planets, being farther off than the earth, they, seen
from Mercury, appear much smaller than to us, and afford very
little light to that planet.

I don't feel so much for its inhabitants on that account, replied
the Marchioness, as from the inconvenience they must suffer from
such excessive heat. Let us try if we can't relieve them in some
way. Is it not probable they have long and plentiful showers, such
as we are told fall continually for four months together, in our hot
countries, at the seasons when the heat is most intense?

It may be the case, answered I; and we may have another way
of giving them relief. There are some parts of China which, from
their situation, ought to be very hot, and yet, even in the month
of July and August, the weather is so cold, that their rivers freeze.
This coldness arises from the quantities of salt-petre with which
the countries abound; the exhalations, drawn up in great abun-
dance by the heat, are of a cold nature. Mercury, if you please,
shall be a little planet made of salt-petre, and the sun, by attracting
the cooling exhalations, will thus prevent the evil it would otherwise be the cause of. However, we may rest assured that nature
would not place beings where it was impossible for them to exist;
and that habit, and ignorance of a better climate, render this situation agreeable: Mercury therefore may perhaps do very well without salt-petre, or abundant rains.

After Mercury, you know, we find the sun. We cannot possibly
place inhabitants there: the *why not* fails in this case. We conclude

from the earth being inhabited, that other bodies of the same nature muſt be so too: but the sun does not resemble the earth, and the reſt of the planets. He is the source of all that light which the planets only refleĉt to each other after they have received it from him. They make exchanges, if I may so express myself, with one another, but none of them can beſtow an original light. The sun is the sole proprietor of that treasure; which he diſtributes freely on every side. The light, thus issuing from the centre, is reflected from every solid body it meets, and from one planet to another, it proceeds in bright ſtreams that intermix, and cross each other in a thousand direĉtions, forming a splendid tissue of the rich- eſt materials. The grand luminary, by being placed in the centre, is in the moſt advantageous situation for animating each planet with his heat and radiance. The sun, then, is of a peculiar nature, but what that nature is, we find it difficult to imagine. Formerly it was believed to be a pure fire, but lately we have been undeceived by observing spots on the surface. As certain new planets had juſt before been discovered, (I shall give you an account of these planets hereafter;) which entirely engrossed the attention of the philosophers, a sort of mania for new planets seized their minds, and they immediately concluded these spots were some; that they performed a circle round the sun, and necessarily concealed some part of his light, by turning their dark side towards the earth. The learned already, through these planets, complimented the different princes of Europe. Some gave them the name of one prince; some of another, and perhaps in time there would have been a great conteſt to know who had the beſt right to name these spots.

I don't like their plan, said the Marchioness. You told me, the other day, that the different parts of the moon were named after learned men; I thought that very proper; as princes monopolize the earth, it is but fair that aſtronomers should have the sky for their share, and not suffer princes to intrude on their domain. Allow them, however, I replied, if territory should be wanting, to consign to them some planet, or some part of the moon. As to these spots on the sun, they can make no use of them; for inſtead of planets, we find they are only clouds of smoke or dross arising from the sun. Sometimes these clouds are greatly accumulated, sometimes we see little of them, and at other times they totally disappear. Sometimes a number of them are combined together, then they are separated into small parts; at one time they are very dark, at another they grow pale. It appears as if the sun was some kind of liquid: many people think it is melted gold, in a continual ſtate of ebullition, producing impurities, which the rapidity of its motion caſts up from the surface; they are afterwards consumed, and others produced. Only think what amazing bodies these are. Some of them are seventeen hundred times[*][71] larger than the earth, for you muſt know, the earth is more than a million times smaller than the sun.[†] Imagine, therefore, what muſt be the quantity of this liquid gold, or the extent of this ocean of light and fire.

Other philosophers say, and with great plausibility, that the spots, or at leaſt the greateſt part of them, are not newly produced, and then deſtroyed after a certain time; but large, solid masses, of irregular forms, always subsiſting; sometimes floating on the surface of the sun, sometimes partly, or entirely buried in the liq-uid subſtance, and presenting to our view different projeċtions ac-cording to the size of the part that remains uncovered. Perhaps they may be parts of some great mass of matter which serves as aliment to the fire of the sun. However, let the sun be what it will, it does not by any means appear habitable.[‡][72] It is a pity; the situation would be advantageous: placed at the centre, its inhabitants would

* The largeſt of the sun's spots are scarcely three times larger than the diameter of the earth, or twenty-seven times its bulk.

† The earth is only a hundred, or to speak with more exaċtness, a hundred and eleven times, smaller.

‡ Some natural philosophers have, however, thought that the sun might be the cause of heat without being itself hot: and that there was a possibility of its being inhabited. M. Herschel believes its population very abundant. *Trans. Philos.* 1795. *Decade Philos-ophique.*

see the planets going round them in regular orbits, whilſt to us their motions seem to have perplexing varieties, which are merely the effeĉt of our not observing them from the beſt place: that is, the centre of their circles. What a sad thing it is: there is but one spot where the ſtudy of the celeſtial bodies would be extremely easy, and at that spot there is nobody to pursue the ſtudy. You forget yourself, answered the Marchioness. Were any one placed on the sun, he would neither see the planets nor the fixed ſtars; would not the light of the sun efface every other objeĉt? The inhabitants would doubtless think themselves the only people in exiſtence.

I acknowledge my error, I replied: I was thinking of the situation of the sun, without considering the effeĉt of such an excessive light; but although you have so properly correĉted my miſtake, yet you muſt allow me to tell you that you have fallen into one yourself. The inhabitants of the sun would not see any: they would be either incapable of enduring so immoderate a light, or, were their eyes sufficiently ſtrong, of receiving it, unless they were at some diſtance; therefore the sun could only be a habitation for people without sight. In short, we have abundant proofs that this luminary was not intended to be a dwelling-place; and therefore we may as well continue our planetary journey. We are now ſtopping at the central point, which is always the loweſt part in anything that is round; and, by the way, I should tell you that in going from our world to this centre, we have travelled thirty-three millions of leagues. We muſt now return the way we came. We pass by Mercury, Venus, the Earth, and the Moon, all which we have visited. Then we arrive at Mars. I don't know that there is any thing remarkable of this planet. The days there are about half an hour longer than ours; and the years twice the length of ours, except a month and a half. Mars is four times less than the earth,* and the sun appears rather smaller and less brilliant than it does to us. In short, Mars contains nothing calculated to arreſt our attention.

*   Its volume, or bulk, is five times smaller.

But what a beautiful object is Jupiter, surrounded by his four moons, or satellites![73] These moons are four little planets which, whilst Jupiter revolves in twelve years round the sun, constantly go round him, as the moon does round the earth. But, interrupted the Marchioness, how is it that there are planets which go round other planets, no better than themselves? It seems to me, that there would be much more regularity and uniformity in assigning to all the planets but one sort of orbit, in which they should move round the sun.

Ah! Madam, I replied, were you but acquainted with the vortices of Descartes; those vortices, so terrible in name, and so charming in the ideas they give rise to; you would not talk in this way. My wits must all go, said she laughing. I must know what these vortices are. Make me quite mad at once: now I have dipped into philosophy, I can't trouble myself about the care of my senses; spite of the world's laughter, we will talk of the vortices. I did not know you had so much enthusiasm, said I; 'tis pity it has no other object than vortices.

What we call a vortex is a quantity of matter, whose detached parts move all in the same direction, but allowed at the same time to have some little movements peculiar to themselves, provided they still pursue the general course. A vortex of wind, for instance, is a vast number of little particles of air, turning all together in a circular direction, and involving whatever comes in their way. The planets, you know, are borne along by the celestial fluid, which is prodigiously subtle and active. All the celestial matter, from the sun to the fixed stars, constantly turns round, carries the planets along with it, and makes them proceed round the sun in the same direction, but in longer or shorter periods, according to their distance from the centre. Even the sun is made to turn on its axis by being exactly in the midst of this moving matter; you will therefore observe, that if the earth were in the central situation she could not be exempted from this rotation.

Such is the great vortex of which the sun is master; but the planets, at the same time, form little vortices in imitation of the sun. Each of them, while turning round the sun, turns likewise on itself, and carries in its motion a certain portion of the celestial matter, which is ready to receive any impulse that would not prevent it from following the general course: this is a vortex of any particular planet, and it extends as far as the motion of this planet has any influence. If a smaller planet comes within the vortex of a larger one, it is irresistibly carried round that larger one, and altogether, the large and small planet, and the vortex that encloses them, perform their revolution round the sun. Thus at the commencement of creation we obliged the moon to follow us, because she came within the influence of our vortex, and was by that means subjugated to our will. Jupiter, the planet we were speaking of, was more fortunate, or more powerful than the earth. Four little planets were in his neighbourhood, and he became master of them all; and we, who are a planet of some importance, would probably have felt his power, if we had been near him. He is a thousand times larger than the earth;* and would easily have drawn us into his vortex, and made us one of his moons; instead of this we have a planet to attend on us, so true is it, that the situation into which we are thrown decides the fate of our lives.

And how do we know, answered the Marchioness, that we shall always remain where we are? I begin to tremble lest we should be foolish enough to approach such an enterprising planet as Jupiter, or that he should come to us, for the sake of drawing us into his vortex; for I can't help thinking, from your description of the agitated state of this celestial fluid, that it must move the planets irregularly, sometimes urging them nearer together, sometimes sending them to a greater distance. We may as well expect to gain as to lose by such an eccentric motion, said I; perhaps we may make a conquest of Mercury or Mars, which are smaller planets,

---

*   We may even say thirteen hundred times.

and incapable of resisting us. However we have no occasion for
either hope or fear; the planets will remain in their places; and,
like the former kings of China, they are forbidden to aim at conquest. You have observed that when oil is mixed with water, the
oil swims at the top. Put any substance that is extremely light on
both these, and the oil will support it, so that it shall not touch
the water: but put a heavier body, or a certain weight, it will pass
through the oil, which is too weak to stop it, and keep falling till it
meets the water, which has sufficient force to bear it up. Thus two
liquors put together, being of unequal weight, will not mix; but
place themselves in different situations; and neither will one rise,
nor the other descend: pour on these other liquors which are of a
nature to remain separate, and the same effect is still produced. In
like manner the celestial matter which fills this grand vortex, is in
separate strata, encircling each other, and of unequal weights, like
oil and water, and some other liquors. Some planets likewise are
heavier than others,* each therefore stops in the layer which has
the degree of force necessary for supporting it, and keeping it in a
state of equilibrium; and you must be convinced that it can never
go beyond this stratum.

I understand, replied the Marchioness, that the different degrees of weight are sufficient to keep them in their proper ranks.
I wish with all my heart there was some such regulating power
among us, that would serve to fix people in the situation most suitable to them! You have quite removed my uneasiness with regard
to Jupiter. I am very glad he will let us remain quietly with our
little vortex, and single moon. I feel very well contented with one
attendant, and do not envy him his four.

You would do wrong if you did, said I; he has no more than
are necessary. He is five times farther from the sun than we, that
is, a hundred and sixty-five† millions of leagues distant from it,

* The Cartesians carried their illusion so far as to believe that so solid a mass as a
planet could be steadily supported by the etherial fluid, the most subtle of all fluids.

† Calculated with more exactness, 179.

consequently his moons receive and reflect but a feeble light: the number, therefore, compensates for the little effect produced by each: were they not separately so inefficient, four moons would appear unnecessary, as Jupiter turns on his axis in ten hours, and, of course, the nights are very short. The satellite which is nearest to Jupiter, performs its circle round him in two-and-forty hours; the next in three days and a half; the third, in seven; the fourth, in seventeen; and by the inequality of their progress, they form a most pleasing spectacle for this planet. At one time they rise all four together; then, almost immediately separate, sometimes they are all at the fall, placed in a line, one above another; afterwards they are seen at equal distances in the sky; then, when two are rising, the other two will set. Above all I should like to see the perpetual variety of eclipses among them, for there is not a day passes in which they do not eclipse each other, or the sun.* Surely as eclipses are so familiar to the inhabitants of that world, they must be considered a subject of amusement rather than terror, as they are here.

You will not fail, I suppose, said the Marchioness, to people these four moons, though they are only little subaltern planets, intended merely to give light to another during the night. Undoubtedly not, I replied. These little planets are not unworthy of inhabitants, because they are unfortunate enough to be subjected to a larger planet.

I think, then, answered she, these satellites ought to be like colonies to Jupiter; that their inhabitants should, if possible, receive from him their laws and customs, and in return, render him some degree of homage, and always consider the great planet with respect. Would it not be needful, said I, for the moons occasionally to send deputies to Jupiter, who should take an oath of fidelity to him? I must own the little superiority we possess over the people in our moon makes me doubt whether Jupiter has much influence

---

* Or, we may add, in which they are not eclipsed by the shadow of Jupiter, which happens the most frequently.

over the inhabitants of his satellites, and I think the only superiority he can aspire to is that of impressing them with awe. For of
what a terrific size he muſt appear! To the planets neareſt to him he
looks sixteen hundred times larger than our moon appears to us.[*]
Truly if the Gauls in ancient times were afraid the heavens would
fall and crush them to death, the inhabitants of this moon may
with greater propriety aprehend the fall of Jupiter. Perhaps, she
replied, that is the subjeċt of alarm to them inſtead of the eclipses,
which you assure me they see without fear;[†] for as they are exempt
from one folly, they muſt be subjeċt to some other. Undoubtedly,
answered I. The inventor of a third syſtem, which I mentioned
the other day, the celebrated Tycho Brahe, one of the greateſt
aſtronomers that ever lived, felt none of the vulgar terror at an
eclipse; he was too much accuſtomed to ſtudy the nature of such
a phenomenon: but what do you think he was afraid of inſtead? If
when he firſt went out of doors the firſt person he saw was an old
woman; or if a hare crossed the path he had taken, Tycho Brahe
thought the day would be unfortunate, and returning in haſte to
his apartment, he shut himself up without venturing to engage in
any occupation whatever.

It would be unjuſt, said she, if such a man as he could not with
impunity overcome the fear of an eclipse, for the inhabitants of
the satellite we were speaking of, to be exempted from it on easier
terms. We will not spare them: they shall submit to the general
doom; and if they escape one error, they shall be liable to another.

A difficulty has juſt occurred to me, continued she, you muſt
remove it if you can: if the earth is so small in comparison of

* Thirty-six times larger than we see the moon: and they receive from him one thousand two hundred and ninety times more light.

† Their solar eclipses are of much longer duration than others.

Jupiter, are we visible to the inhabitants of that planet? I am afraid we are unknown to them.*

Really I think so, answered I; the earth is certainly too small to be diſtinguished by them.†

We can only hope that in Jupiter there may be some aſtronomers who, after taking great pains to compose very excellent telescopes, and availing themselves of the fineſt nights for making their observations, may at length discover a very little planet, which they had never seen before. At firſt the learned give an account of it in their journal; the reſt of the people either hear nothing about it, or laugh at it when they do; the philosophers are discouraged, and resolve not to mention it again, and but a few of the inhabitants, who are more reasonable than the others, will admit the idea: By and by they examine again; they see the little planet a second time; they are then assured of its reality, and even begin to think it has a motion round the sun. After observing it a thousand times, they find out that this revolution is performed in a year: and at laſt, when the learned have been at great pains to inveſtigate the subjeđ, the inhabitants of Jupiter know that our world is in the universe. The curious eagerly look through their telescopes, and, with all their looking, can scarcely discern it.

Were it not disagreeable, said she, to know that from Jupiter we can only be seen through telescopes, I should amuse myself with the idea of all the glasses being pointed towards the earth, as ours are towards him, and the mutual curiosity with which the two planets examine each other, and enquire, *What world is that? What sort of people inhabit it?*

Your imagination is too rapid, I replied; when the aſtronomers of Jupiter become acquainted with our earth, they do not become

*   The earth at that diſtance muſt appear only three seconds and a half in diameter, as the planet Herschel [Uranus] does to us; but our nearness to the sun necessarily prevents them from seeing us at all.

†   The earth at that diſtance muſt appear only three seconds and a half in diameter, as the planet Herschel does to us; but our nearness to the sun necessarily prevents them from seeing us at all.

acquainted with us: they will not suspect the possibility of its being inhabited; if anyone should venture to express such an idea,
how they would laugh at him! Perhaps they would even perse-
cute any philosopher who should maintain the opinion. After all,
I think the inhabitants of Jupiter are too much occupied in making
discoveries on their own globe, to concern themselves about us.
Jupiter is of such extent, that if they are adepts in navigation, their
Christopher Columbus must be fully employed. The inhabitants
cannot know, even by reputation, a hundredth part of the other
inhabitants. In Mercury, on the contrary, they are all neighbours,
living familiarly together, and hardly considering the tour of their
world more than a pleasant walk. If we are not visible to Jupiter,
much less can Venus be so, who is at a still greater distance;* and
Mercury must be most out of its reach of all, being the smallest,
and the most distant. However, the inhabitants can see Mars, their
own four satellites, and Saturn with all its moons. Surely then they
have planets enough to perplex their astronomers: nature, in kindness, has hid from them the rest.

What! cried the Marchioness, do you consider it a kindness?
Without doubt, answered I. This great vortex contains sixteen
planets; nature to spare us the trouble of studying the motions of
so many, lets us see but seven: is not that a favour? But not feeling
the value of this mark of consideration, we have, with great pains,
discovered the other nine, which had been concealed from us: our
curiosity brings its own punishment, in the laborious study which
astronomy now requires.

I see, she replied, by the number of planets you mention, that
Saturn must have five moons† You are right, said I: and it is but
just that he should have so many; as he is thirty years in going
round the sun; and in some parts the night lasts fifteen years, for
the same reason that on our globe, which turns in a year, there are

---

* Venus is not farther from Jupiter, but more concealed by the rays of the sun.

† He has seven, and Herschel six. In all there are twenty-five planets, without reckon-
ing ninety-one comets known in 1800.

nights, beneath the poles, of six months' duration. But Saturn, being at twice the diſtance that Jupiter is from the sun, consequently
ten times farther than the earth, his five moons, faintly as they are
illumined, would not give sufficient light during his nights; he has
therefore a wonderful resource, the only one of the kind we have
discovered in the universe: 'tis a large circle or ring* which environs the planet, and which, being sufficiently elevated to escape
almoſt entirely the shadow of Saturn, refleĉts the sun's light on
the darkened parts, and refleĉts it more ſtrongly than all the five
moons, because it is not so high as the loweſt of them.

Really, said the Marchioness, with an air of deep refleĉtion and
aſtonishment, all this is managed with wonderful order; nature had
certainly in these inſtances a view to the wants of living beings:
this admirable disposition of light was not the effeĉt of chance.
Only the planets which are diſtant from the sun have been provided with moons—the earth, Jupiter, and Saturn; for Venus did not
require any; nor Mercury who already has too much light; whose
nights are extremely short, and probably considered a greater
blessing than even the days. But ſtop—I think Mars, who is farther from the sun than we, is without a moon. We cannot conceal
the faĉt, I replied; he has none; but he doubtless has resources for
the night which we are ignorant of. You have seen phosphorus;
matters of that kind, whether liquid or dry, receive and imbibe
the light from the sun, which they emit with some force when in
the dark. Mars perhaps has high rocks of phosphorus that absorb,
in the daytime, light enough to irradiate the night. You muſt own
it would be an agreeable sight for the rocks to light up as soon as
the sun was set, and without art, produce the moſt magnificent
illuminations, that with all their radiance, would not have the inconvenience of caſting any heat. In America, you know, there are
birds which in the dark will afford light enough to read by;[74] how
can we tell whether Mars has not a great number of such birds,

---

*   Its exterior diameter is sixty-seven thousand seven hundred leagues.

who, as soon as the night is come, disperse themselves on every side, and give an artificial day?

I am not satisfied, answered she, either with your rocks or birds. They would be pretty enough to be sure; but as nature has bestowed so many moons on Saturn and Jupiter, it shews that moons are necessary. I should have been very much pleased to find that all the worlds at a great distance from the sun had some, if Mars had not formed a disagreeable exception. Ah! replied I, if you were more deeply versed in philosophy you must accustom yourself to see exceptions to the best systems. We clearly see that some things are adapted in the most perfect manner to their end; others we accommodate as well as we can, or perhaps are obliged to content ourselves with knowing nothing about them. Let us do so with respect to Mars, since our researches are fruitless, and resolve to say no more about him.

We should be very much surprised, were we on Saturn, to see during the night a great ring, extending over our heads in a semi-circular form from one end of the horizon to the other; and by reflecting the light of the sun, would have the effect of a moon at every part of the circle. And are we not to have inhabitants in this great ring? said she, laughing. Though I am disposed to place them wherever I can, answered I, I confess I dare not tell you there are any there; this ring appears too irregular a dwelling. As for the five moons, we can't dispense with inhabitants for them. If the ring, however, were what some suppose, only a circle of moons, following each other very closely, with an equal motion, and the satellites, five of these moons escaped out of the ring, what numbers of worlds would the vortex of Saturn contain! Be that as it may, the people in Saturn are uncomfortable enough, even with the help of their ring. It gives them light, it is true; but what sort of light, at that immense distance from the sun? The sun himself, which appears to them a hundred times smaller* than to us, seems

---

* Ten times less in diameter.

but a little pale star, emitting but a feeble light or heat. And could they be transported to our coldest countries, such as Greenland and Lapland, you would see them ready to expire with the heat. If water were conveyed to their planet it would no longer be water, but a polished stone, and spirits of wine which never freeze here, would become hard as diamond.

Your description of Saturn petrifies me, said the Marchioness; though just now you almost threw me into a fever in talking of Mercury. The worlds, answered I, which are at the different extremities of an immense vortex, must be totally unlike.

Then, replied she, the people are very wise in Saturn, for you told me they were all mad in Mercury. If they are not very wise, answered I, they are at least, I suppose, very phlegmatic. Their features could not accommodate themselves to a smile; they require a day's consideration before they answer any question, and they would think Cato of Utica[75] unmanly and frivolous.

I am thinking, said she, that all the inhabitants of Saturn are slow; all those of Mercury are quick; amongst us some belong to the former class, some to the latter; may not that be in consequence of the earth's being placed just in the middle situation and participating of both extremes? The men of our world have no determined character; some are like the inhabitants of Mercury; others resemble those of Saturn, in short we are a compound of all the other planets. That's a good idea, replied I; we form such a ludicrous assemblage that it might easily be imagined we had been brought together from a variety of worlds. We are therefore very well situated for studying character, for this is an abstract of all the planets.

At any rate, rejoined the Marchioness, the situation of our world has one great convenience: the heat is not oppressive as at Mercury or Venus, nor the cold so benumbing as at Jupiter or Saturn. And we are in a part of the earth that is not subject to the greatest degrees of heat and cold experienced even on our own globe. If a certain philosopher returned thanks to his Creator for having formed him a man, and not a beaſt; a Greek and not a Barbarian; I think we ought to be grateful for being born on the moſt temperate planet in the universe, and in one of the moſt temperate parts of that planet. You ought likewise, madam, said I, to be thankful for being young, and not old; young and handsome, not young and ugly; a young and handsome French woman, not a young and handsome Italian: there are many things to excite your gratitude besides the temperature of your climate.

Ah! replied she, let us be grateful for every thing, even the vortex in which we are placed. The happiness we enjoy is but little, we muſt not lose any of it; it is well to cultivate an intereſt in the moſt common things. If we are only alive to ſtrong emotions our pleasures will be few, seldom attainable, and dearly purchased. Promise me, then said I, that when such animated pleasures are within your reach you will think of the vortices and me, and not negleĉt us entirely. Very well, said she: but will philosophy always afford me new enjoyments? For to-morrow, at leaſt, answered I: I have the fixed ſtars in reserve for you, which surpass all that you have yet examined.

FIFTH EVENING

*Every Fixed Star is a Sun, which
Diffuses Light to its Surrounding Worlds*

THE MARCHIONESS WAS very impatient to know what the fixed stars were. Are they inhabited, like the planets? said she, or are they not peopled? What can we make of them? Perhaps you would find out what they are, answered I, if you were to try. The fixed stars cannot be at less distance from the earth, than twenty-seven thousand, six hundred and sixty times* the earth's distance from the sun, which is thirty-three millions of leagues; perhaps some astronomers would tell you they are farther still. The space between the sun and Saturn, the most distant planet, is only three hundred and thirty millions of leagues; that is but a trifle in comparison of the distance between the sun, or earth, and the fixed stars, in fact, we don't take the trouble to compute it. Their light, as you perceive, is brilliant: if they receive it from the sun, it must be very faint after travelling such an immense journey, and by reflecting it to us it would be still more weakened. It would be impossible for light, which had twice gone such a long space, to appear so bright as that of the fixed stars. They are therefore luminous in their nature, or in other words, they are so many suns.

Do I mistake, cried the Marchioness, or do I see your drift? Are you not going to say 'the fixed stars are all suns': our sun is the centre of a vortex which turns around him; why should not each fixed star be also the centre of a vortex, turning round it? Our sun enlightens planets; why should not every fixed star likewise

* Or even two hundred thousand times.

enlighten planets! I need make no other answer, replied I, than Phœdrus made to Enone: *thou hast named it.*

But, rejoined she, you are making the universe so unbounded that I feel lost in it; I don't know where I am, nor what I'm about. What! are they all vortices heaped in confusion on one another? Is every fixed star the centre of a vortex, as large perhaps as ours?* The amazing space comprehending our sun and planets is but a little portion of the universe! An equal space, occupied by each of these vortices? The thought is fearful; overwhelming! For my part, said I, I think it very pleasing. Were the sky only a blue arch to which the stars were fixed, the universe would seem narrow and confined; there would not be room to breathe: now that we attribute an infinitely greater extent and depth to this blue firmament, by dividing it into thousands of vortices, I seem to be more at liberty; to live in a freer air; and nature appears with astonishingly increased magnificence. Creation is boundless in treasures; lavish in endowments. How grand the idea of this immense number of vortices, the middle of each occupied by a sun, encompassed with planets which turn round him! The inhabitants of one of these numberless vortices, on every side behold the suns of surrounding vortices, although the planets belonging to them are invisible, as the light they receive from their suns cannot penetrate beyond their own vortex.

You are directing my eye, answered she, to an interminable perspective. I see, plainly enough, the inhabitants of the earth; then you enable me to discover, with somewhat less clearness, those of the moon and other planets contained in our vortex. After all that you require me to view the people that dwell in planets belonging to other vortices. I must confess they are so much in the back ground that with all my efforts they are scarcely perceptible to me. In short do they not seem almost annihilated by the very expression you are obliged to make use of

---

* That may be the case, but we have no proof that there are planets turning round these stars.

in describing them? You muſt call them the inhabitants of one
of the planets, contained in one, out of the infinity of vortices.
Surely the very idea of ourselves is as nothing when such a description is applied to us, when we are thus loſt amongſt millions
of worlds. For my part, the earth begins to diminish into such a
speck, that in future I shall hardly consider any objeƈt worthy
of eager pursuit. Surely people, who form unnumbered schemes
of aggrandizement, who are wearing themselves out in following up projeƈts of ambition, are ignorant of the vortices. I think
my augmentation of knowledge will encrease my idleness, and
when I am reproached for being indolent I shall reply, *Ah! if you
knew the hiſtory of the fixed ſtars!* Alexander could not have been
acquainted with it, answered I; for a certain author, who believes
that the moon is inhabited, tells us very seriously that it was impossible for Ariſtotle to avoid receiving so rational an opinion
(could Ariſtotle be ignorant of any truth), but that he never dis-
closed it for fear of displeasing Alexander, who would have been
miserable to hear of a world that he could not subjugate. There
would have been a ſtill greater reason for keeping the vortices
of the fixed ſtars a secret; if any body in those days had known
them, they would not have thought of ingratiating themselves
with the monarch by talking of them. It is unfortunate that I who
am acquainted with the syſtem should not be able to reap any
benefit from it. According to your reasoning it will only be an
antidote to the disquietudes of ambition; that is not my malady.
The weakness I am moſt addiƈted to is an excessive admiration
of beauty, and I fear the vortices will have no power to assiſt me
in overcoming it. The immense number of worlds deſtroys the
grandeur of this, but it does not lessen the charms of a fine pair
of eyes or a beautiful mouth; *they* retain their power in spite of
all the worlds that can be created.

Love is a ſtrange thing, said she, laughing; it escapes every cor-
reƈtive; there is no syſtem that can abate its influence. But answer

me seriously; have you sufficient reason for believing this system? To me it appears to rest on an uncertain foundation. A fixed star is of a luminous nature like the sun, therefore you say it must, like the sun, be a centre to a vortex containing planets which travel round the sun. Now, is that a necessary consequence?

Listen, madam, I replied; we are so naturally disposed to mingle the follies of gallantry with our gravest discussions, that mathematical reasoning partakes of the nature of love. Grant ever so little to a lover, and presently you are forced to grant him a great deal more, and so on till you don't know how to stop. In like manner admit any principle a mathematician proposes, he then draws a consequence which you are obliged to admit, and from that consequence another, and thus before you are aware he carries you so far, that on a sudden you wonder where you have got to: these two characters always take more than you mean to give them. You must own that when two things are similar in all that I know of them, I may reasonably think them similar in what I am unacquainted with in respect to them. From that principle I draw the conclusion of the moon being inhabited because she resembles the earth; and the other planets, because they resemble the moon. And because the fixed stars bear a resemblance to our sun, I attribute to them all that he possesses. You have already made too many concessions to draw back, you must go on; do it therefore with a good grace. But, said she, in admitting this resemblance between the fixed stars and our sun, we must suppose that the inhabitants of another great vortex see it as a little fixed star, visible only during their nights.

That is indisputable, I replied. Our sun is so near to us in comparison of the suns belonging to other vortices that his light must be incomparably stronger to us than to them. When he is risen we can discern no other heavenly body: so, in another vortex, another sun eclipses ours, and permits it to appear only at night, with all the other suns, then visible. With them, fixed to the blue firmament, our sun forms a part of some imaginary figure. As to the planets that go round him, as they are not seen at so great a distance; they are not so much as thought of. Thus all the suns are daily luminaries to their own vortex, and nightly ones to all the other vortices. Each reigns alone in his own system; elsewhere, is but one of a great number. Nevertheless, she asked, do not these worlds differ from each other in a thousand instances, notwithstanding this equality? for a general resemblance does not exclude a vast number of dissimilarities.

Surely, answered I: but the difficulty is to find them out. For aught we know one vortex may have more planets revolving round its sun, another fewer. In one there are subaltern planets, turning round the principal planets; in another they may be all alike. Here they all collect round their sun in a circle, beyond which is an empty space which extends to the neighbouring vortices; in other parts of the universe they may have their orbits at the extremities of their vortex whilst the centre is left empty. And very likely there are some vortices without any planets; others, whose suns, not being in the centre, have a circular revolution, carrying their planets along with them; others, again, whose planets may rise and set with regard to their sun according to the change of that equilibrium which keeps them suspended——What would you have more? Surely here is enough for a person who has never been beyond one vortex.

All that is nothing, she replied, for the number of worlds. What you have been imagining would suffice but for five or six, instead of millions.

If you talk of millions now, said I, how will you count them when I tell you there are many more fixed stars than you discover; that with telescopes, an endless number are seen which are invisible to the naked eye; and that in a single constellation, where we might before have counted a dozen or fifteen, there have been found as many as we were accustomed to observe throughout the heavens?*

Have pity on me, cried she; I yield; you have overwhelmed me with worlds and vortices. Ah! said I, but I must add something more still; you see that white part of the sky, called the milky-way. Can you guess what it is?——An infinity of little stars, invisible to our eyes on account of their smallness, and placed so close to each other that they seem but a stream of light. I wish I had a telescope here to shew you this cluster of worlds. In some measure, they resemble the Maldivia Isles, those twelve thousand little islands[76] or banks of sand, separated only by narrow canals of the sea which one might almost leap over. The little vortices of the milky-way must be so close, that from one world to another the people might converse or shake hands. The birds, at least, I think, can go from one world to another; and pigeons may be taught to carry letters as they do in our Levant from one town to another. These little worlds must deviate from the general rule by which the sun of any vortex effaces, at its rising, all the other suns. In one of the little vortices contained in the milky-way the sun of that particular vortex can hardly appear closer to its planets, or more brilliant, than a hundred thousand other suns, in the neighbouring vortices. The sky, then, is filled with countless quantity of fires almost close to each other. When they lose sight of their own sun, they have thousands still remaining; and the night is not less enlightened than the day; at least the difference is so trifling that we may say there is no night. The inhabitants of those worlds, accustomed as they are to perpetual light, would be very much astonished to hear of miserable creatures who spend half their time in profound darkness; and who, even during the light of day, see but one sun.

---

* I conclude, from a pretty accurate calculation, that we may perceive a hundred millions with a telescope that has an opening of four feet; I have clearly distinguished fifty thousand, and my glass is but two inches and a half in diameter.

They would think we had fallen under the displeasure of nature, and shudder at our condition.

I don't ask you, said the Marchioness, whether they have any moons in the milky-way; they could be of no use to the principal planets, since they have no nights, and besides that, move in so small a space that they could not be encumbered with subaltern planets. But continued she, by multiplying worlds so liberally, you give rise to a great difficulty. The vortices, of which we see the suns, touch our vortex: the vortices you say are round; can so many circles touch this single one? I can't understand how it is.

It shews a great deal of sense, answered I, to discover this difficulty, and even to be unable to solve it, for it is in itself well founded, and in the way you conceive it, unanswerable; therefore there would be but little proof of wisdom in finding an answer to what was incapable of any. If our vortex were in the figure of a die, it would have six flat sides, which is very far from a circle; on each of these sides might then be placed a vortex of the same shape. If instead of six it had twenty, fifty, or a thousand, flat sides, an equal number of vortices might come in contact with it, each resting against one of these sides. You know the greater number of flat sides a body has, the nearer it approaches in form to a circle; so that a diamond cut into a great number of facets, if they were extremely small, would be nearly as round as a pearl of the same size. The vortices are only circular in this manner. They have an amazing number of flat sides, each of which is close to another vortex. These sides are very unequal; some larger, some smaller. The smallest correspond to those of the milky-way. If two vortices leave any space between, which must often be the case, nature, to make the most of the extent, fills up the vacancy by one or two, or perhaps a thousand, little vortices, which without incommoding any of the others, form one, two, or a thousand more systems of worlds; so there may be many more worlds than our vortex has sides; and I dare say, though these little vortices are formed merely to fill up spare corners of the universe that would otherwise have been useless; though they may be over-looked by the neighbouring vortices, yet they are quite satisfied with themselves. It is probably such little vortices whose suns we cannot discover without telescopes, of which there is a prodigious number. In short all these vortices are adjusted in the best order imaginable; and as each of them must turn round its sun without changing place, it is formed to move in the most easy and commodious manner for that purpose. They, as it were, catch hold of each other, like the wheels of a watch, and mutually assist the motion. It is likewise true that in a sense they counteract one another: each vortex if it had no external pressure would extend itself; but when it attempts to swell, it is repelled by the surrounding vortices, which forces it to shrink back; then it extends again, and so on:* some philosophers think that the fixed stars give such a sparkling intermittent light in consequence of this alternate expansion and contraction of the vortices.

There is something agreeable, said the Marchioness, in the idea of such a combat among the worlds, and the reciprocal emission of light produced by it, which apparently is the only communication carried on between them.

No, no, I replied, that is not the only one. The neighbouring worlds sometimes send us visitors, who come in a very magnificent style. These visitors are comets,† ornamented with brilliant flowing hair, a venerable beard, or a majestic train.

Ah! what ambassadors! said she, laughing. We could dispense with their company, for they only frighten us. They only frighten children, answered I, because their appearance is extraordinary; but there are many children among us. The comets are merely planets, belonging to another system. Their orbit was towards the extremity of their vortex, which was perhaps differently com-

* The preservation of the starry system is more satisfactorily explained by attraction; they are all kept in equillibrium by their mutual attraction.

† It is indisputably proved that the comets belong to our solar system.

pressed by those that surrounded it: the lower side, on that account was flatter than the top, and the lower side was next to us. These planets, beginning at the upper part to form their circle, did not foresee that it would extend beyond the limit of their vortex, at the lower part; in order, therefore, to continue their circular journey, they were obliged to enter the extremities of the next vortex, which we will suppose is ours. They always appear to us extremely elevated, moving on the other side of Saturn. Considering the prodigious distance of the fixed stars, there must be between Saturn and the extremities of our vortex a great space void of planets. Our enemies reproach us with the inutility of this space, but we find there is a use for it, as it is devoted to the service of foreign planets, that occasionally enter our system.

I understand, said she; we don't allow them to penetrate into the heart of our vortex, and mix with our planets; we receive them as the Grand Seignior receives the ambassadors that are sent to him. He does not honour them with a lodging in Constantinople, but assigns them one in the environs. There is another resemblance, I replied, between us and the Ottomans; they receive ambassadors without sending any in return; and we receive the comets without sending any of our planets to return their visits.

From all these circumstances, answered she; we seem to be very proud; yet we should not hastily form that conclusion; these strange planets have a very menacing air with their beards and trains; perhaps they are only sent to insult us; ours not having so imposing an appearance would not be so well calculated to inspire those worlds with awe. The tails and beards, I replied, are merely extraneous: the planets themselves do not differ from ours; but in entering our vortex they assume the beard or train from a certain illumination derived from our sun. This, by the bye, has not been very well explained by our astronomers; however, they are sure it is only some sort of illumination, and they must tell us more of it when they can. Then I wish, rejoined she, that our Saturn would take a beard or a tail, and frighten the other vortices; then laying aside his terrific appendages, return to us, and perform his ordinary functions. He would do better to stay where he is, answered I. You recollect I explained to you the shock produced by the repulsive power of each vortex: I think a poor planet must be violently shaken in such a situation, and the inhabitants cannot feel much the better for passing through it. We think ourselves vastly unfortunate when a comet makes its appearance, whereas we ought to consider the comet most unfortunate. I am not inclined to pity it, said the Marchioness; I dare say all its inhabitants arrive here in good health, and it must be extremely entertaining to them to enter into a new vortex. We who always remain in our own have but a dull life. If the people in a comet have the sense to know the time at which they shall pass into our vortex, those who have already been on the journey, are just before busily employed in describing to the rest what they will see. Speaking of Saturn, they say: 'You will presently see a planet with a great ring round it. Then, you will discover one followed by four small planets.' Some of these people, perhaps, are set to watch the moment of entering our system: when it is arrived, they cry, *new sun, new sun,* as our sailors exclaim, *land, land.*

I find then, said I, it is useless to attempt raising your compassion for the comets: I hope, however, you will not refuse it to the inhabitants of a vortex, whose sun has been extinguished, and who are thus condemned to perpetual darkness. Suns extinguished! cried she. Yes, undoubtedly, I replied. The ancients saw certain fixed stars, which are no longer visible.* These suns have been deprived of their light; ruin must have ensued throughout the vortex; a general mortality on all the planets; for how could existence be maintained without the sun? The thought is too dreadful, said she; is it not possible to evade it? I'll tell you, answered I, what some very intelligent people have imagined.

* In 1572 and 1604, some beautiful stars appeared to burst into light, and afterwards became extinct. *Astron.*, Art. 792.

They think that the fixed stars that have disappeared are not extinct, but partly darkened; that is to say, that they have one side
obscure; the other luminous: that as they turn on themselves, they
first present the light part to us, and then the dark; when that is the
case, we cease to see them. Apparently the fifth moon belonging
to Saturn is in this condition, for during one part of its revolution
we entirely lose sight of it; at which time it is not most remote
from the earth; on the contrary, it is then sometimes nearer than
when visible. Though the moon is a planet, and therefore cannot
exactly guide our opinion with respect to suns, yet we may suppose that a sun can be partly covered with fixed spots. To spare
you the pain of believing the other opinion, we will adopt this,
which is more agreeable: but I can only receive it when applied to
such fixed stars as have a regular time for appearing and disap-
pearing, as some have lately been observed to do, otherwise we
cannot suppose them half suns. What must we say to the stars that
disappear, and do not become visible after a time that would certainly have been sufficient for turning on their axis ? You are too
just to require me to believe that they are half suns: however I will
do all in my power to serve you; we will conclude that these stars
are not extinguished, but plunged in the unfathomable depth of
the sky, and thus become invisible; in this case the vortex would
accompany its sun, and all go on as usual. It is true that the greatest part of the fixed stars have not any motion which removes them
farther from us, for if they were not always equally distant, they
would sometimes appear larger, sometimes smaller; but that is not
the case. We will therefore suppose that some of the small vortices
being light and active, slip between the others, and return after
they have made their tour, whilst the larger systems remain immoveable. But there is one inevitable misfortune: there are some
fixed stars, which for a long time are alternately visible and invisible, and at length totally disappear. Half suns would reappear at
a regular time; others that had retreated to an immense distance,

would at once disappear, and be concealed for a very long time: exert therefore all your resolution, Madam; these ſtars are certainly suns which grow so dark as to be invisible to us, then resume their brightness, and afterwards are entirely extinguished. How, exclaimed the Marchioness, can a sun, a source of light become darkened? With the greateſt ease, answered I, if Descartes be in the right. He imagines that the spots on our sun, being impurities, or vapours, may grow thick, colleĉt together, form themselves into a mass, and continue to encruſt the sun till it is quite hid. If the sun is a fire conneĉted with solid matter, serving as its aliment, we are not in a better condition: the solid matter may be consumed. 'Tis said we have already had a fortunate escape: the sun during several years (the year, for inſtance, after the death of Caesar), appeared very pale; owing to the incruſtation which was beginning to form. The sun had sufficient force to break and disperse it; had it continued, we should have been loſt. You make me tremble, said the Marchioness. Now I know the consequences of paleness in the sun, inſtead of going to my glass every morning to see if *I* am pale, I think I shall go and look whether *the sun* is so. Take courage, Madam, I replied, it requires a good deal of time to ruin a syſtem of worlds. But, answered she, it seems as if time would inevitably effeĉt it. I cannot take upon me to deny it, said I. The immense mass of matter which composes the universe is in continual motion, even the smalleſt particles of it, and since there is this motion we are in danger, for changes muſt happen, either slowly or rapidly, but always in a time proportioned to the effeĉt. The ancients were so vaſtly wise as to imagine the heavenly bodies were of such a nature as never to alter, because they had not observed any alteration in them. Had they leisure to assure themselves of this by experience? Compared with us the ancients were young: if flowers that laſt but a day were to transmit their hiſtories to each other, the firſt would draw the resemblance of their gardener in a certain way: after fifteen thousand ages of these flowers had elapsed, others would still describe him in the same manner. They would say, 'We have always had the same gardener, the memoirs composed by our ancestors prove this to be the case; all their representations exactly apply to him; surely he is not mortal like us; no change will ever take place in him.' Would the reasoning of these flowers be conclusive?—it would have a better foundation than that of the ancients respecting the celestial bodies; and had there never to this day been observed any change in the heavens, though they should appear likely to remain much longer without alteration, I would not yet decide on them; I should think more experience necessary. Should the term of our existence, which is but a moment, be the measure of other durations? Ought we to assert that what has lasted a hundred thousand times longer than we, must last for ever? No, ages on ages of our duration would scarcely be any indication of immortality. Truly, said the Marchioness, I think these worlds can have no pretensions to it. I shall not do them the honour to compare them with the gardener who outlives so many transient flowers; they are but as those flowers themselves, springing up and fading away, one after another: for I suppose, if old stars disappear, new ones become visible; the species cannot otherwise be continued. Yes, answered I, we need not fear the extinction of the species. Some will tell you these new stars are only suns which re-approach us after having been for a long time at a distant part of the heavens. Others think they are suns that have broken through the crust that began to cover them. I easily conceive the possibility of all this; but I think it equally possible for new suns to be created. Why should not the matter that is fit to compose a sun, after having been dispersed in various places, be at length gathered together in one spot and then become the foundation of a new system of worlds? I am the more inclined to this opinion because it answers better to the grand idea I entertain of the works of nature. Has she now no way of producing and destroying plants and animals but by a continual revolution? I am persuaded, and I doubt not that by this time you are so too, that she exerts the same power with respect to the worlds. But on such subjects we can only form conjectures. The fact is that for nearly a century past, in which, by the help of telescopes, almost a new heaven has been discovered, unknown to the ancients, there have been few of the constellations in which some sensible alteration has not taken place;* the greatest number of changes is observed in the milky-way, as if more motion and bustle existed among this heap of worlds. Really, said the Marchioness, I find the worlds, in short all the heavenly bodies, so liable to change that I have quite overcome the horror I felt at the idea of the sun's being extinguished. Well, replied I, to prevent you from relapsing, we will say no more about them; we are arrived at the uppermost part of the heavens, and to inform you whether there are any stars beyond that, exceeds my skill. You may place more worlds or not, just as you are disposed. These invisible countries should, in propriety, be left to the philosophers: they may imagine them to exist, or not exist or to exist in any way they chuse. I shall content myself with having directed your mind to all that is discernible by your sight.

Ah! she exclaimed, then I am acquainted with the whole system of the universe! how learned I am! Yes, said I, you are learned enough in all reason, and your knowledge is attended with this convenience,—you may extract your belief of all I have told you whenever you think proper. I only ask as a reward for my trouble, that whenever you see the sun, the sky, and the stars, you will think on me.†

---

* This is not proved.

† As I have given these conversations to the public, I think it would not be right to conceal any thing which passed on the subject. I shall publish another dialogue of the same kind that we had a long while after these. It shall be entitled the 'Sixth Evening,' as the rest were evening scenes.

Sixth Evening

Additional Thoughts in Confirmation of
Those in the Preceding Conversations.
Discoveries that have been lately made in
the Heavens

For a long time the Marchioness and I said nothing about the plurality of worlds; we had apparently forgotten that we had ever talked on the subject. I went one day to her house, and just as I entered, two men of talents and celebrity were going out. You see, said she, what visitors I have had; I assure you they are gone away with a suspicion that you have turned my brain. I should be very proud of such an achievement, answered I; it would shew my power, for I think one could not devise a more difficult undertaking. Well, replied she, I am afraid you have accomplished it. I don't know how it happened, but whilst my two friends, whom you met at the door, were here, the conversation turned on the plurality of worlds; perhaps they had an invidious design in directing it to that subject. I immediately told them all the planets were inhabited. One of them said he was certain I could not be of that opinion: in the most unaffected manner, I maintained my sincerity; he continued to think I was only feigning, and I believe he had too great a regard for me to admit the possibility of my having really adopted so extravagant an opinion. The other, from esteeming me less, did not doubt my veracity. Why have you made me obstinately adhere to sentiments which people who have the greatest friendship for me will not suffer themselves to believe me possessed with? But, madam, answered I, why did you maintain these opinions seriously, when talking with persons that I am sure would not gravely argue on any subject? Should we thus trifle with the inhabitants of the planets? Let us, who believe their existence, be content to remain a little select band, and not

disclose our mysteries to the vulgar. Vulgar! exclaimed she; do
you reckon those two men among the vulgar? They have good
understandings, said I; but they never reason. Grave reasoners,
who are austere people, would not hesitate to place them in that
class. They however take their revenge by ridiculing the reasoners. We should if possible accommodate ourselves to persons of
both characters; it would have been better to speak jestingly of
the planetary inhabitants to such men as your two friends, since
they are accustomed to pleasantry, than to enter on an argument,
for which they have no talents. You would have retained their
good opinion without depriving the planets of a single inhabitant. Would you have meanly sacrificed the truth? answered she.
Where is your conscience? I must own, I replied, I have not much
zeal for truths of this nature; I would readily forbear to maintain
them if it suited my convenience.

The cause which prevents people from believing the planets
to be inhabited is, that they appear to them only bodies placed in
the heavens to give light, instead of globes consisting of meadows and fruitful countries. We readily believe that meadows and
fields are inhabited, but it is thought ridiculous to assert that
mere luminous bodies are. 'Tis in vain that reason informs us of
fields in the planets; reason comes too late, the first coup d'œil
has impressed our minds before-hand, and this impression is not
willingly parted with. The planets, 'tis said, are only luminous
bodies; what sort of inhabitants then can they have? Our imaginations do not enable us to distinguish their figures, therefore it is
the shortest way to deny their existence. Would you require me,
for the sake of establishing the idea of these inhabitants, whose
interest cannot be very dear to me, to attack all the powers of
the senses and the imagination? Such an enterprize would de-
mand a vast deal of courage. Men are not easily persuaded to see
through their reason, rather than their eyes. Some few persons
are rational enough to believe, after a thousand proofs have been

given them, that the planets are worlds like ours, but they do not believe it in the same way they would do, if they had not seen them apparently so different; they always recur to the first idea they had formed, and can never wholly divest themselves of it. These people seem to *condescend* to our opinion, and only patronize it from a love of singularity.

Is not that enough, said she; for an opinion that is merely probable? You would be astonished, answered I, if I told you the word probable was too modest for the occasion. Is it merely probable that Alexander has been in existence? No, you consider it certain; and what is the ground of your certainty? Is it not that you have had every proof that such a subject requires, and that no circumstance leads you to doubt the fact? You have never seen Alexander, nor have you any mathematical demonstration of his existence. What would you say if this were the case with respect to the inhabitants of the planets? We cannot shew them to you, nor can you require us to demonstrate their being, in a mathematical way; but you have all the evidence that can be desired: the entire resemblance between the planets and the inhabited earth; the impossibility of imagining any other use for which they could be created; the fruitfulness and magnificence of nature; the attention she seems to have paid to the wants of their inhabitants, such as giving moons to those planets that were very remote from the sun, and the greatest number of moons to the most distant: and it is an important consideration that every thing is on that side of the question, without any objections to counterbalance it; you cannot for a moment doubt unless you resume the vulgar mode of seeing and thinking. In fact it is impossible to have more evidence, and evidence of a more determinate kind; how then can you treat this opinion as a mere probability? But do you think, said she, I can feel as certain that the planets are inhabited, as that Alexander has been in existence? By no means, I replied; for although, on the subject we are speaking of we have as many proofs as in our situation we can receive, yet these proofs are not numerous. I protest, exclaimed she, I'll renounce these planetary inhabitants, for I don't know whether to believe there are any or not—it is not certain, yet it is more than possible—I am quite perplexed. Do not be discouraged, madam, I replied, Clocks that are made in the most common manner shew the hour; those only that are made with more exquisite art, indicate the minutes; in like manner common minds see a great difference between probability and absolute certainty; but it is only superior understandings that ascertain the degrees of certainty or of probability, and who, if I may use the expression, can tell the minutes as well as the hours. Place the inhabitants of the planets a little below Alexander in point of certainty, but above a vast number of historical relations which are not entirely proved; I think that is their proper place. I love order, said she, you do me a kindness in giving arrangement to my ideas: why did you not do this before? Because, answered I, whether you admit to this idea a little more, or a little less certainty than it possesses is not of much consequence. I am certain you do not feel so assured as you ought to do of the earth's motion: are you the less happy on that account? Oh! as to that opinion, I am sure I do my duty; you have no right to complain of me, for I firmly believe that the earth turns. Yet I have not given you the most convincing proof of it, answered I. You use me very ill, said she, to make one believe things without sufficient reason; am I unworthy to hear the best arguments? I wished to prove my opinions I replied, by easy, entertaining arguments; would you have had me make use of such solid, sturdy ones as I should have attacked a doctor with? Certainly, said she; now fancy me a doctor, and let me have this new proof of the earth's motion.

With all my heart, answered I; it is this, and I am vastly pleased with it, because I think I found it out myself: but it is so good and so natural that I can hardly hope to have been the inventor. I am sure an obstinate learned man who wished to oppose it would be forced to talk a great deal on the occasion; and that is the only way in which a scholar can be overcome. It is evident either that all the heavenly bodies go round the earth in four-and-twenty hours, or the earth, turning on her axis, only imagines the motion in them. It is the moſt improbable thing in the world that they should in reality go round the earth in that short space of time, though we are not as firſt aware of the absurdity of such an opinion. All the planets certainly revolve round the sun: but these revolutions are unequal from the unequal diſtances at which they are placed from the sun: the moſt remote, as we might naturally suppose, take a longer time than the reſt. This order is observed even in the satellites that go round a large planet. Jupiter's four moons, and the five belonging to Saturn, require a longer or shorter time to move round their planet according to their diſtance from it. It is further ascertained that the planets have a rotation on their own axis; the time of this is likewise unequal; we cannot tell the cause of such inequality, whether it depends on the different size, or the degree of solidity of the planets, or on the different degrees of rapidity of the vortices in which they are enclosed, and the liquid matter by which they are carried along;* this inequality however is certain, and in general we find that the order of nature is such as to admit of particular variations in things that are regulated by the same rules.

I underſtand, said the Marchioness; I am quite of your opinion; if the planets moved round the earth, the time employed by each would be different, according to their various diſtances, as is the case in their revolutions round the sun: is not that what you mean? Precisely so, madam, answered I; their unequal diſtances from the earth would produce an inequality in their revolutions round her: and the fixed ſtars, being so extremely remote from us, so far beyond all that could have a general movement round us, at leaſt situated in a place where such a motion muſt be very feeble, is there any

* We can assign no reason; the irregularity depends on the original cause, whatever that cause may be, which at firſt determined their motion.

probability of their revolving round us in four-and-twenty hours, like the moon which is so near to us? Ought not the comets likewise which do not belong to our vortex, which have such irregular courses, and such different degrees of swiftness, to be exempt from performing this daily circle round our world? No, planets, fixed ſtars, and comets too, muſt all turn round the earth! Were there but a few minutes difference in the time of their revolutions we might be satisfied with it; but they are all exaĉtly equal never varying in the slighteſt degree; surely this is a suspicious circumſtance.

Oh! replied the Marchioness, I could venture to say this exaĉtitude exiſted only in our imaginations. I am glad that any thing inconsiſtent with the genius of nature, which this equality in so many moving bodies would be, should depend on our motion, and she, even at our expence, be free from the charge of inconsiſtency. For my part, said I, I dislike a perfeĉt regularity, and I don't approve of the earth's turning every day on her axis in exaĉtly twenty-four hours; I am disposed to think the time varies. Varies! she exclaimed; do not our clocks shew that it is always equal? Oh; replied I, I don't depend on clocks, they cannot always be perfeĉtly right; and should they be so, and sometimes shew that the earth has made a longer or shorter tour in four-and-twenty hours than usual, it would be thought that we ought rather to suspeĉt them of being wrong than to attribute any irregularity to the revolutions of the earth. That is paying an extravagant respeĉt to her, I should depend no more on the earth than on a clock: the one might be put out of sorts almoſt by the same causes as the other, only I think it would take longer time to produce a sensible irregularity in the earth; that is the only advantage I should allow her to have over a clock. Might not the earth by degrees get nearer to the sun, and then, finding herself in a situation where the matter was agitated with greater violence, perform her motion on her axis, and her revolution round the sun, in a shorter time? In that case the years would be shorter, and the days too, but we should not perceive the difference, for we should still divide the year into three hundred and sixty-five days, and the days into twenty-four hours. So that without living longer than we do now, we should live a greater number of years: and on the contrary, if the earth were to remove farther from the sun, we should live fewer years, although our lives would be as long. In all probability, said she, if that were possible, a long succession of ages would make but a trifling difference. True, I replied; nature does nothing abruptly, her method is to effect every alteration by such gentle gradations that it is scarcely perceptible to us. We hardly observe even the changes of seasons; others that are produced much more slowly must in general escape our notice. Nevertheless every thing is subject to mutability; even a certain lady who has been seen, through telescopes, in the moon for about forty years, appears considerably older. She used to be rather handsome: now her cheeks are fallen away, her nose and chin are beginning to meet; in short all her charms are fled, and it is even feared that her life is near its close.[*]

What are you talking of! cried the Marchioness. I am not jesting, I replied. A figure has been observed in the moon which resembled a woman's head rising from among the rocks, and in that part an alteration is perceived. Some pieces have fallen off a mountain, and left the points which appear like the forehead, nose and chin of an old woman. Does it not seem, said she, as if some malignant power had a spite against beauty, since the young lady's head is the only spot in the moon that has undergone a change. Perhaps, answered I, to make amends, the alterations on our globe may give additional beauty to some face observed by the inhabitants of the moon, I mean some face formed like those of the people in that planet, for we always try to discover in distant objects, the resemblance of what we continually think of. Our astronomers discern young ladies' faces in the moon; probably if women were

[*]   We are not assured that this alteration has taken place in the part of the moon that has some resemblance to a woman's head; but there must be changes, if we judge by the volcano which has been repeatedly observed. *Astron.* Art. 3339.

to examine it they would find handsome male faces. If *I* were to look, I don't know whether I would not see your likeness madam. I muſt undoubtedly, said she, feel myself obliged to any body who could find me there; but let us return to what we were talking of juſt now; are there any considerable alterations on the earth? In all probability there are, answered I. Many high mountains, at a great diſtance from the sea, have on them beds of shells, which shew that they were formerly covered with water. Sometimes likewise, at a diſtance from the sea, are found ſtones containing petrified fishes. How could they have got to that place unless the water had been there? Fables tell us, that Hercules separated with his hand two mountains called Calpe and Abila, which being situated between Africa and Spain obſtructed the ocean; and the sea immediately rushed in violently, and formed the great gulph that we call the Mediterranean. Fables are not altogether fabulous; they are hiſtories of remote periods disguised by two very ancient and common defects; ignorance, and a love of the marvellous. It is not very credible that some Hercules (for there have been fifty), the ocean may have torn asunder, perhap with the assiſtance of an earthquake two mountains more feeble than the reſt and have by that means rushed in between Europe and Africa. Then a new spot was discovered on our globe, by the people in the moon, for you recollect, madam, that the water forms a dark spot. It is the general opinion that Sicily has been separated from Italy, and Cyprus from Syria: new islands have sometimes been formed in the sea; earthquakes have ingulfed some mountains, and produced others, as well as changed the course of rivers. Philosophers give us reason to fear that the kingdom of Naples and Sicily, being over great subterranean vaults filled with sulphur, will sometime or other fall in, when the vaults are no longer ſtrong enough to resiſt the fires contained in them, which now have vent at such openings as Vesuvius and Ætna. All this will be sufficient to diversify a little the appearance we make to the inhabitants of the moon.

I would rather tire them, said the Marchioness, with a monoto-
nous appearance, than entertain them in the ruin of provinces.

That is nothing, answered I, to what takes place in Jupiter. He
appears to be surrounded with belts, which are diſtinguished from
each other; or from the spaces betwixt them, by their different degrees of light. These are land and seas, or at leaſt parts of the planet
differing in their nature.[77] Sometimes these lands grow narrower,
sometimes wider. New ones are formed in various parts, and
some of the old ones disappear: and all these changes visible only
through our beſt telescopes, are in themselves much more considerable than if our oceans were to inundate all the land, and leave
its own bed to form new continents. Unless the inhabitants of Jupiter are amphibious, and live with equal ease either on land or
water, I hardly know what can become of them.* We see likewise
great alterations on the surface of Mars, even from one month to
another. In that short time seas overflow large continents, and retire by a flux and reflux a thousand times more violent than ours:
or if this be not the case, some change equivalent to it takes place.
Our planet is very quiet compared with these; we have great reason to congratulate ourselves, especially if it is true that in Jupiter
countries as extensive as Europe have been set on fire. Set on fire!
cried the Marchioness; that would he a great piece of news there.
It would indeed, answered I. We have observed in Jupiter, for perhaps twenty years, a long ſtream of light more brilliant than the
reſt of the planets.† We have had deluges here, but very seldom:
perhaps in Jupiter they have now and then a large conflagration as
well as frequent deluges. But be that as it may, the brilliant light I
spoke of is very different from another, which apparently is as old
as the world, though it has but lately been discovered.‡ How can a

* These lands surrounding Jupiter, which are sometimes in great numbers, are apparently clouds.

† I don't know that this observation is authentic.

‡ The zodiacal light. *Aſtron.*, Art. 844.

light be formed for concealment? said she, that is something quite out of the common way.

This light, I replied, is only visible at twilight, which it most frequently long enough, and of sufficient power to conceal it; and when it is not hid by the twilight, either the vapours of the horizon prevent us from seeing it, or without great attention we may even mistake it for twilight. However, about thirty years since it was discovered with certainty, and for some time gave great light to the astronomers, whose curiosity wanted stimulating by something new. They might find as many new subaltern planets as they chose without feeling any interest in them. The two last moons of Saturn, for instance, did not enrapture them as Jupiter's satellites had done; custom destroys the power of every thing.

We see, during a month before and after the equinox of March, when the sun is set and the twilight disappeared, a sort of whitish light resembling the tail of a comet. It is seen before the dawn and sun-rise, towards the equinox of September, and morning and night towards the winter solstice. At other times, as I have before said, the twilight conceals it; for we have reason to believe it always exists. It has lately been conjectured that it is produced by a large mass of matter, somewhat dense, which environs the sun for a certain extent. The greatest part of his rays penetrate this covering, and come to us in a straight line but some of these rays by striking against the internal surface are reflected back to us, either before the direct rays can reach us in the morning, or after they have ceased to enlighten us in the evening. As these reflected rays come from a higher region than the direct ones, it is therefore earlier when we receive them, and later before we lose them.

On this ground I must retract what I said on the probability of the moon having no twilight, for want of a surrounding atmosphere as dense as that of the earth. She is no loser by it, if she can receive a twilight through this thick air which surrounds the sun, and reflects his light to places which could not have his direct rays. Then enquired the Marchioness, will not this be a source of twilight to all the planets, without the necessity of a dense atmosphere to environ each, since that which surrounds the sun may produce the same effect for all the planets in the vortex? From the frugality of nature, I am disposed to believe she has effected the purpose by this mean only. Yet, said I, in spite of this frugality, the earth would have two causes of twilight, one of which (the dense air before the sun), would be useless, and could only serve as an object of curiosity to the frequenters of the observatory: but it may be that the earth alone sends out exhalations sufficiently gross to produce twilight; and therefore a general resource has been provided for the other planets; if their evaporations are more pure and subtle. We perhaps, of the inhabitants of all the worlds in our vortex breathe the grossest air; did the people of the other planets know that, with what contempt they would survey us!

That would be wrong, answered the Marchioness; we are not contemptible for being surrounded by a thick atmosphere since the sun himself is in the same situation. Tell me, is not this air produced by certain vapours that you formerly told me issued from the sun; and may it not be to moderate the power of the first rays which perhaps would otherwise be excessive? I think it probable that the sun may be thus veiled, to accommodate it to our use. That is a happy idea, madam said I; you have founded a pretty little system. We may add that this vapour possibly falls back in a sort of rain to refresh the sun, in the same manner as we sometimes throw water into a forge when the fire becomes too fierce. We cannot attribute too much to the power of nature; but all her operations are not made visible to us, therefore we cannot feel assured of having discovered her designs, or her manner of acting. We should not consider any new discovery a certain foundation for reasoning on, though we are very much inclined to do it: philosophers are like elephants, that in walking never put one foot to the ground till they feel the other firmly supported. That comparison, said she, is the moſt juſt because the merit either of elephants or philosophers does not consiſt in external charms; we shall however do well to imitate the superior judgment of both: inform me more of the new discoveries, and I promise not to be in a hurry again to form syſtems.

I have told you, I replied, all the news I have heard from the sky, and I believe no later intelligence has been received. I am sorry it is not so entertaining and wonderful as some observations I read the other day in an abridgment of the *Annals of China,* written in Latin. They there see a thousand ſtars at a time fall from the sky into the ocean with an amazing noise; or dissolve and disperse in rain. This has not merely been seen once in China; I have met with the same account given at two remote periods of time, besides that of a ſtar which goes towards the eaſt, and burſts with the noise of a gun. It is a pity such sights should be confined to China, while this part of the world is never favoured with them. It is not long since all our philosophers thought they had had sufficient experience to pronounce the heavens and all the celeſtial bodies incorruptible and incapable of change; and at the same time people at the other end of the world were seeing ſtars dissolve by thousands: their appearance muſt have differed very much from ours. But, said she, I never heard that the Chinese were great astronomers. No, answered I, but the Chinese are gainers by being at so great a diſtance from us, as the Greeks and Romans were by being separated by a long space of time; whatever is remote assumes the right of imposing on us.

Really I am more and more of opinion that Europe is in possession of a degree of genius which has never extended to any other part of the globe, at leaſt not to any diſtant part. It is not perhaps able to diffuse itself over a great proportion of the earth at once and some invincible fatality prescribes to it very narrow bounds. Let us then make use of it while it is in our possession: and let us rejoice that it is not confined to science and dry speculations, but equally extended to objects of taste, in which I doubt whether any people can equal us. Such, madam, are the things that should engage your attention and constitute your philosophy.

The End

Notes

1. H. A. Hargreaves in his translation of *Plurality of Worlds* for the University of California Press (1990) refers to an 1801 edition of Gunning's translation, but I have been unable to trace it in any library. Nor is there any sign of it having been advertised. Given the Gunning family's taste for publicity, I suspect that the 1801 edition is a chimera.

2. [Nicholas Malebranche], *Father Malebranche His Treatise Concerning the Search after Truth*, T. Taylor trans., London, 1700, p.64.

3. Patricia Phillips, *The Scientific Lady: A Social History of Women's Scientific Interests 1520–1918*, p. 47.

4. Alain Niderst, *Fontenelle à la Recherche de Lui-Même*, Nizet, Paris, 1972.

5. A verse translation of a fable by Father Commire, one of his teachers.

6. The *Mercure Galant* was founded by Jean Donneau de Visé (1638–1710) in 1672. It was a gazette and literary magazine, containing poetry, anecdotes and articles on court life, fashion and plays. It is thought to have had a large female readership.

7. *The Works of M. de Voltaire*, translated by T. Smollett, T. Francklin, et al., London, 1761, p.

8. The letters are largely *galantes* in the sense of being flirtatious. To Madame de G. he writes, 'For a long time, madame, I would have taken the liberty of loving you, had you the time to be loved by me...' To Mlle de C* who had recently arrived in France from England: 'I am writing to you, mademoiselle, in a language which you do not yet understand well, but in recompense I'll write to you on a matter which you'll have no difficulty understanding. When I say I find you the most lovely person in the world, I don't think you would need an interpreter...' To Mlle De T: 'The terrible news I've heard, mademoiselle! You are going to marry my rival!'

9. The first category includes Anacreon with Aristotle, Homer with Æsop and Candaules with Gyges (Candaules: 'The more I think about it, the more I'm convinced that you didn't have to murder me'). The second category includes a discussion between Socrates and Montaigne, and one between the Greek anatomist Erasistratus of Chios and William Harvey concerning the circulation of the blood. The third category includes Molière with Paracelsus, and Mary Queen of Scots with David Riccio.

10. Some sources give 1687 as the first publication date, but this is wrong. The title page states 'A Paris: Chez G. de Luyne,… En la boutique de la veuve C Blageart,… Et T. Girard, M DC. LXXXVI.' Title page dates are occasionally wrong, but there is no reason to believe that this one is.

11. *The History of Oracles and the Cheats of the Pagan Priests*, London, 1688. It must have been popular because it was reprinted in the following year.

12. Richard Watkins (translator), *Jérôme Lalande, Diary of a Trip to England, 1763*, Kingston, Tasmania, 2002, p. 163, f.n.

13. *Voyage d'un François en Italie, Fait dans les Années 1765 & 1766*, Paris, 1769.

14. *Diary of a Trip to England*, p. 164.

15. *Ibid.*, pp. 162–2.

16. 'Reamur relates, on the authority of M. de la Hire, that a young French lady could never resist the temptation of eating a spider, whenever she met with one in her walks. They are said to taste of nuts, at least this was the opinion of the celebrated Maria Schurrman, who not only ate them, but justified her taste by saying that she was born under Scorpio. Latreille informs us, that the astronomer Lalande was equally fond of this offensive morsel.' – *The Literary Gazette and Journal of Belles Lettres, Arts, Sciences, &c.*, London 1829, p. 694.

17. Lady [Constance] Russell, *Three Generations of Fascinating Women and Other Sketches from Family History*, 1905, p.140.

18. When George II, then an old man, regretted that there were no masques for her entertainment, Maria told him that the only thing she was looking forward to was a coronation.

19. Lady Bury was also a novelist, probably as successful as her cousin Elizabeth in terms of sales, but inferior to her in stylistic terms.

20. Quoted in *Three Generations of Fascinating Women*, pp. 140–141.
21. *Ibid*, p. 139.
22. Although the *Dictionary of National Biography* lists this portrait as being after R. Saunders, I can find no such painter. It was probably by John Sa[u]nders (1750–1825) who appears in the Royal Academy catalogue for 1775 as 'John Saunders junior', his father having also been a painter.
23. Elizabeth Gunning is firing letters marked 'love letters', 'forged Love Letter', 'Letter from Marq: of Bland[ford] written by myself', 'Letter written by my Daddy', 'Letter forged by my Mother', 'Letter forg'd by myself', 'Letters in Answer to myself'. The speech bubbles read as follows: Gunning – 'I find our Stratagem won't take effect, & therefore I'll be off; & manoeuvre; – any common Soldier can lead <u>on</u>, to attack, but it requires the skill of a <u>General</u> to bring off his forces with honor after a defeat –'; ELIZABETH – 'O Mother! Mother! my mask'd Battery is discovered, & we shall be blown up! O Mother, Mother, we must raise the siege immediately, & take refuge under the Duchess's cover'd way, & there act on the defensive: O Mother! Mother! its all your fault, say what you will!'; MRS GUNNING – 'Good Heav'ns! who could have thought that the siege of a coronet would have ended in smoke & stink! – well, I'll take my <u>affidavit</u> that I know nothing at all about the matter'; DUCHESS OF BEDFORD – 'Come under my Protection, deary's! I'll hide you in Bedfordshire; & find one of my little Granny-boys, to play with Missy.
24. *Lloyds Evening Post*, 21 February 1791.
25. Saturday, 7 August 1790.
26. It is rather hard to imagine the Duke of Marlborough being so in awe of the ancient Gunning family, but it accords well with the General's own opinion. He wrote to the *Public Advertiser* on 24 December 1790 and stated in reference to his sister, Elizabeth, 'she was certainly of illustrious descent, who derived her birth in the thirty-third generation from CHARLEMAGNE.'
27. *Lloyds Evening Post*, 21 February 1791.
28. This is presumably the same unscrupulous Captain Essex Bowen of the 82nd regiment of foot whose (unwilling) mistress, Mary Anne Talbot, sailed with him for Santo Domingo in 1792, disguised as a boy and calling herself John Taylor. She served as a drummer-

boy at the battle of Valenciennes, at which Bowen was killed. She continued to work on various warships disguised as a man for some years.

29. John Gunning, *An Apology for the Life of Major General G———*, London, 1792, p. 82.

30. *Ibid.* p.29

31. *Lord Fitzhenry*, 1794, *The Foresters*, 1796; *The Orphans of Snowdon*, 1797; *The Gipsy Countess*, 1799; *The Farmer's Boy*, 1802.

32. *Three Generations of Fascinating Women*, pp. 141–2.

33. *Astronomie des Dames* (1786) in the *Bibliothèque Universelle des Dames* series.

34. The English antiquarian, Richard Chandler (c. 1737–1810), wrote in his *Voyages dans l'Asie Mineure et en Grèce*, vol. 3, Paris, 1803, p. 436, 'On peut compter parmi les Ahténiens modernes éclairés, Mr Codrika, ci-devant premier interprête de l'ambassadeur turc à Paris, qui a fait en grec, moitié vulgaire, moitié ancien, une traduction très-élégante de Mondes de Fontenelle.'

35. Johann Elert Bode (1747-1826), director of the Berlin Observatory. Like Fontenelle, he wrote a popular astronomical text, *Anleitung zur Kenntnis des gestirnten Himmels* (1768). It was Bode who gave the name Uranus to the planet discovered by Sir William Herschel.

36. Nicolas Charles Joseph Trublet (1697–1770) was a canon at St Malo. He was elected to the Académie française in 1761. He was best known for his *Essais sur Divers Sujets de Littérature et de Morale* (1735).

37. The biographical dictionary, *Le grand Dictionaire Historique*, originally published by Louis Moréri (1643–1680) in 1674.

38. Jan Heweliusz, latinised to Hevelius (1611–1687), Polish astronomer and mayor of Gdansk.

39. *Nouvelles de la république des lettres*, edited in the early days by the Huguenot Pierre Bayle (1647–1706). A number of Fontenelle's pieces were published therein.

40. Johann Heinrich Lambert (1728–1777), Swiss astronomer and mathematician.

41. Georges-Louis Leclerc, Comte de Buffon (1707–1788), French mathematician, naturalist and cosmologist.

42. Gowin Knight (c.1713–1772), English physician and supplier of compasses to the Royal Navy, whose *Two Simple Active Principles*,

Attraction and Repulsion (1748) suggested that magnetism, light, heat and gravity were the result of attractive and repulsive particles moving in 'ætherial fluid'.

43. Sir William Herschel (1738–1822), German-born British composer, astronomer and talented designer of telescopes. The correct reference is 'On the Nature and Construction of the Sun and Fixed Stars,' *Phil. Trans. Roy. Soc. London*, vol. 85 (1795), pp. 46–72. Herschel believed that it was the sun's atmosphere which was radiating light and that sun spots were glimpses of its (much cooler) surface. He states (p. 63), 'I think I am authorized, *upon astronomical principles*, to propose the sun as an inhabitable world'.

44. Horace-Bénédict de Saussure (1740–1799), Swiss physicist and alpinist.

45. Jean le Rond d'Alembert (1717–1783), French mathematician and physicist who co-edited the famous *Encyclopédie, ou Dictionnaire Raisonné des Sciences, des Arts et des Métiers* with Denis Diderot.

46. Jerôme de Lalande provides a footnote on Cartesian vortices further on.

47. *La Princesse de Clèves*, a very successful historical novel published anonymously in 1678. It is often attributed to Madame Lafayette. There was a very favourable review in the *Mercure Galant*, one which has been attributed, rightly or wrongly, to Fontenelle.

48. The Plough (UK) or Big Dipper.

49. Babylonian.

50. Piraeus, the port of Athens.

51. 'To awaken a man who is deceived as to his own merit is to do him as bad a turn as that done to the Athenian madman who was happy in believing that all the ships touching at the port belonged to him.' François de la Rochefoucauld, *Maxims*, 93. The madman was Thrasyllus, son of Pythodorus.

52. The definition of a league is either 3 statute miles (4.828 kilometres) or 3 nautical miles (5.556 kilometres).

53. Jesuits reported cannibalism among the Iroquois, but they were not the most impartial observers.

54. Miss Gunning has 'Persian'. That this strange mistake should appear twice is hard to explain. The text of the 1803 edition reads 'Parisian'.

55. Miss Gunning's text (both 1803 and 1808 editions) has 'cannot', but this is clearly wrong from the context, and reference to the French text confirms this.

56. Giovanni Domenico Cassini (1625–1712), Italian astronomer who came to the discipline through a youthful interest in astrology. Although the moon has long been assumed to be completely waterless, recent studies suggest that there may be ice at its poles. NASA will be launching the Lunar Reconnaissance Orbiter in 2009 to establish whether this is the case. Nonetheless, Cassini's supposed rivers were clearly imaginary.

57. Both the the 1803 and 1808 editions have 'Gabileus' but French editions have 'Galilée'.

58. Again, both editions have 'decams' for 'dreams'.

59. A character in Ariosto's *Orlando Furioso*.

60. A supposed document granting the Catholic Church temporal power over Rome and the western empire.

61. Sadly just one of many racial or national stereotypes in this work.

62. First news of the Antipodes reached Europe from the voyage of the Dutch ship *Duyfken* in 1606.

63. *L'Astrée*, a novel by Honoré d'Urfé (1568–1625), tells of the love between the shepherd Céladon and the beautiful Astrée.

64. Fontenelle must have had in mind Henry Power's *Experimental Philosophy* (London, 1664) or, more probably, Robert Hooke's *Micrographia* (London, 1665). Hooke wrote 'Of these small Eels, which are to be found in divers sorts of Vinegar, I have little to add besides their Picture… Taking several of these out of their Pond of Vinegar, by the net of a small piece of filtring Paper, and laying them on a black smooth Glass plate, I found that they could wriggle and winde their body, as much almost as a Snake, which made me doubt, whether they were a kind of Eal or Leech.'

65. Venus's sidereal day (the time it takes to make a complete 360° rotation on its axis) is 243 Earth-days. However, its solar day (the time between one sunrise and the next) is only 116.75 days.

66. The radius of Venus is about 95% that of the Earth.

67. Sylvander was the name of a number of heros, including that of Philip Hart's song, *The False Maid:* 'Sylvander once ye gayest swain…'

Madeleine de Scudéry

68. *Clélie* by Madeleine de Scudéry (1607–1701), featuring a map of Arcadia, in which the topographical features revolve around love.

69. Actually, due to the lack of atmosphere, the surface temperature of Mercury varies enormously between night and day. In the day, the temperature reaches about 700 K (427° C); at night it drops as low as 100 K (–173° C). The daytime temperature is substantially below that required to melt iron, silver or gold.

70. Mercury's sidereal day (see note on Venus) is 58.7 Earth-days in duration and its solar day is 176 Earth-days. Its year is 88 Earth-days.

71. Fontenelle's figure seems too high and Lalande's, given in the footnote, too low. One of the largest sunspots, observed in 1947, was estimated to be forty or fifty times larger than the earth.

72. See p. 5 and related endnote.

73. There are 62 known moons of Jupiter.

74. Although there are no birds with truly luminous properties, there are a number which sport highly reflective areas. It is just about possible that a bird contaminated with luminescent fungus or bacteria might glow in the dark.

75. Cato the Younger was famous for his steadfast resistance to Julius Caesar. Gunning has 'Cato of Atica' in both the 1803 and 1808 editions.

76. There are believed to be between 200 billion and 400 billion stars in the Milky Way, whereas there are officially only 1,192 islets in the Maldives.

77. Jupiter's bands are actually dramatic, fast-moving cloud patterns.

TIGER OF THE STRIPE

Typeset in the United Kingdom
by Tiger of the Stripe
in Monotype Fournier
and LTC Fournier Le Jeune
using Adobe InDesign

Breinigsville, PA USA
20 March 2011
258005BV00002B/2/P

9 781904 799375